CENTRIC 2017

The Tenth International Conference on Advances in Human oriented and Personalized Mechanisms, Technologies, and Services

October 8 - 12, 2017

Athens, Greece

CENTRIC 2017 Editors

Stephan Böhm, RheinMain University of Applied Sciences - Wiesbaden, Germany

Lasse Berntzen, Vestfold University College - Tønsberg, Norway

Florian Volk, Technische Universität Darmstadt, Germany

Printed from e-media with permission by:

Curran Associates, Inc.
57 Morehouse Lane
Red Hook, NY 12571

Some format issues inherent in the e-media version may also appear in this print version.

Printed by Curran Associates, Inc. (2017)

International Academy, Research, and Industry Association (IARIA)
412 Derby Way
Wilmington, DE 19810

Phone: (408) 893-6407
Fax: (408) 527-6351

petre@iaria.org

Additional copies of this publication are available from:

Curran Associates, Inc.
57 Morehouse Lane
Red Hook, NY 12571 USA
Phone: 845-758-0400
Fax: 845-758-2633
Email: curran@proceedings.com
Web: www.proceedings.com

Table of Contents

A Survey of the Effectiveness of Automated Revenue Collection Systems in County Governments in Kenya. A Case Study of Kiambu and Taita Taveta County Governments 1
Margaret Njenga and Joseph Sevilla

Scoring of Machine-Learning Algorithms for Providing User Guidance in Special Purpose Machines 7
Valentin Plenk, Sascha Lang, and Florian Wogenstein

Person Re-identification in Crowded Scenes with Deep Learning 14
Yupeng Wang, Huiyuan Fu, and Shuangqun Li

Dynamic Scrutable User Modeling utilizing Machine Learning 18
Dima S. Mahmoud

Complexity Reduction in Graphs: A User Centric Approach to Graph Exploration 24
Tim Grube, Florian Volk, Max Muhlhauser, Suhas Bhairav, Vinay Sachidananda, and Yuval Elovici

Business Process Models as a Foundation for High-Fidelity-Prototypes of Mobile Enterprise Applications 32
Tobias Braumann, David Broll, Matthias Jurisch, and Bodo Igler

The Attitude of German Web Users Toward Foreign Websites for Buying Products Online 36
Bastian Eine and Werner Quint

Influence of the Perceived Data Security and Trust on Usage Frequency of Internet Service 44
Erik Massarczyk and Peter Winzer

Quantifying Mobile User Experience Status quo, Implementation Challenges, and Research Agenda 53
Maria Lusky and Stephan Boehm

Hedonic Motivation of Chatbot Usage - A Wizard-of-Oz Study based on Face Analysis and User Self-Assessment 59
Judith Eisser and Stephan Bohm

CENTRIC 2017

Forward

The Tenth International Conference on Advances in Human-oriented and Personalized Mechanisms, Technologies, and Services (CENTRIC 2017), held on October 8 - 12, 2017- Athens, Greece, addressed topics on human-oriented and personalized mechanisms, technologies, and services, commonly known as I-centric.

There is a cohort of technologies that favored the so called "user-centric" services and applications. While some of them reached some maturity, others are to prove their economics (WiMax, IPTV, RFID, etc). The human-oriented and personalized technologies and services rely on a key set of features, some to be deployed, others getting more mature (personal profiles, preferences, identity, proximity, personal devices, etc.). Following, advanced applications covering human related activities benefit from personalized and human-oriented networks and services, especially preventive and personalized medicine, body networks and devices, or anticipative systems.

The conference provided a forum where researchers were able to present recent research results and new research problems and directions related to them. The conference sought contributions presenting novel result and future research in all aspects of user-centric mechanisms, technologies, and services.

Similar to the previous editions, this event continued to be very competitive in its selection process and very well perceived by the international community. As such, it attracted excellent contributions and active participation from all over the world. We were very pleased to receive a large amount of top quality contributions.

We take here the opportunity to warmly thank all the members of the CENTRIC 2017 technical program committee as well as the numerous reviewers. The creation of such a broad and high quality conference program would not have been possible without their involvement. We also kindly thank all the authors that dedicated much of their time and efforts to contribute to the CENTRIC 2017. We truly believe that thanks to all these efforts, the final conference program consists of top quality contributions.

This event could also not have been a reality without the support of many individuals, organizations and sponsors. We also gratefully thank the members of the CENTRIC 2017 organizing committee for their help in handling the logistics and for their work that is making this professional meeting a success.

We hope the CENTRIC 2017 was a successful international forum for the exchange of ideas and results between academia and industry and to promote further progress in personalization research. We also hope Athens provided a pleasant environment during the conference and everyone saved some time for exploring this beautiful historic city.

CENTRIC Steering Committee

Lasse Berntzen, University College of Southeast, Norway
Stephan Böhm, RheinMain University of Applied Sciences - Wiesbaden, Germany
Mu-Chun Su, National Central University, Taiwan
Florian Volk, Technische Universität Darmstadt, Germany
Nitin Agarwal, University of Arkansas at Little Rock, USA

CENTRIC Industry/Research Advisory Committee

Lisa Yu, BOSCH Research, USA
Vitaly Klyuev, University of Aizu, Japan
Anbang Xu, IBM Research - Almaden, USA
Sabine Coquillart, INRIA, France
Thomas Marx, University of Applied Sciences Bingen, Germany

CENTRIC 2017

Committee

CENTRIC Steering Committee

Lasse Berntzen, University College of Southeast, Norway
Stephan Böhm, RheinMain University of Applied Sciences - Wiesbaden, Germany
Mu-Chun Su, National Central University, Taiwan
Florian Volk, Technische Universität Darmstadt, Germany
Nitin Agarwal, University of Arkansas at Little Rock, USA

CENTRIC Industry/Research Advisory Committee

Lisa Yu, BOSCH Research, USA
Vitaly Klyuev, University of Aizu, Japan
Anbang Xu, IBM Research - Almaden, USA
Sabine Coquillart, INRIA, France
Thomas Marx, University of Applied Sciences Bingen, Germany

CENTRIC 2017 Technical Program Committee

Nitin Agarwal, University of Arkansas at Little Rock, USA
Mohammed A. Akour, Yarmouk University, Jordan
Mehdi Ammi, University of Paris-Saclay - CNRS/LIMSI, France
Angeliki Antoniou, University of Peloponnese, Greece
Lasse Berntzen, University College of Southeast, Norway
Stephan Böhm, RheinMain University of Applied Sciences - Wiesbaden, Germany
Marco Bruti, Telecom Italia Sparkle S.p.A., Rome, Italy
Daniel B.-W. Chen, Monash University, Australia
Sabine Coquillart, INRIA, France
José Luis Cuevas Ruíz, Instituto Federal de Telecomunicaciones, Mexico
Yoanna Dosouto, Alaska Airlines, USA
Thierry Duval, Lab-STICC | IMT Atlantique, France
Rainer Falk, Siemens AG, Germany
Fabio Gasparetti, Roma Tre University, Italy
Huan Gui, University of Illinois at Urbana-Champaign, USA
Riccardo Guidotti, KDD Lab | ISTI-CNR, Pisa, Italy
Till Halbach, Norwegian Computing Center, Norway
Muhammad Iftikhar Hanif, International Medical School | Management & Science University, Malaysia
Qiang He, Swinburne University of Technology, Australia
Tzung-Pei Hong, National University of Kaohsiung, Taiwan
Imène Jraidi, University of Montreal, Canada
Parves Kamal, St. Cloud State University, USA
Saman Kamran, University of Lugano, Switzerland
Teemu Kanstren, VTT, Finland
Vitaly Klyuev, University of Aizu, Japan

Hristo Koshutanski, Safe Society Labs, Spain
Jaroslaw Kozlak, AGH University of Science and Technology, Poland
Daniela Marghitu, Auburn University, USA
Célia Martinie, IRIT, University Toulouse 3 Paul Sabatier, France
Erik Massarczyk, University of Applied Sciences – RheinMain - Wiesbaden Rüsselsheim, Germany
Jose Manuel Molina Lopez, Universidad Carlos III de Madrid, Spain
Thomas Marx, University of Applied Sciences Bingen, Germany
Yoshimasa Ohmoto, Kyoto University, Japan
Valentim Realinho, Polytechnic Institute of Portalegre, Portugal
Ann Reddipogu, RKode Ltd, UK
Michele Risi, University of Salerno, Italy
Luis E. C. Rocha, Karolinska Institutet, Sweden / Université de Namur, Belgium
Elad Michael Schiller, Chalmers University of Technology, Göteborg, Sweden
Gian Luca Scoccia, Gran Sasso Science Institute (GSSI), L'Aquila, Italy
Monica M. L. Sebillo, University of Salerno, Italy
Jingbo Shang, University of Illinois, Urbana-Champain, USA
Jungpil Shin, University of Aizu, Japan
Roman Shtykh, Yahoo Japan Corporation, Japan
Jacques Simonin, IMT Atlantique, France
Christophe Strobbe, Hochschule der Medien, Stuttgart, Germany
Mu-Chun Su, National Central University, Taiwan
Jörg Thomaschewski, University of Applied Sciences Emden/Leer, Germany
Yuan Tian, University of Texas at Dallas, USA
Tsuda Toshitaka, Waseda University / Fujitsu Limited, Japan
Jacques Traoré, Orange Labs, France
Carlos Travieso González, University of Las Palmas de Gran Canaria, Spain
Florian Volk, Technische Universität Darmstadt, Germany
Benjamin Weyers, RWTH Aachen University, Germany
Sanjaya Wijeratne, Kno.e.sis Center | Wright State University, USA
Anbang Xu, IBM Research - Almaden, USA
Lisa Yu, BOSCH Research, USA
Sherali Zeadally, University of Kentucky, USA

Copyright Information

For your reference, this is the text governing the copyright release for material published by IARIA.

The copyright release is a transfer of publication rights, which allows IARIA and its partners to drive the dissemination of the published material. This allows IARIA to give articles increased visibility via distribution, inclusion in libraries, and arrangements for submission to indexes.

I, the undersigned, declare that the article is original, and that I represent the authors of this article in the copyright release matters. If this work has been done as work-for-hire, I have obtained all necessary clearances to execute a copyright release. I hereby irrevocably transfer exclusive copyright for this material to IARIA. I give IARIA permission or reproduce the work in any media format such as, but not limited to, print, digital, or electronic. I give IARIA permission to distribute the materials without restriction to any institutions or individuals. I give IARIA permission to submit the work for inclusion in article repositories as IARIA sees fit.

I, the undersigned, declare that to the best of my knowledge, the article is does not contain libelous or otherwise unlawful contents or invading the right of privacy or infringing on a proprietary right.

Following the copyright release, any circulated version of the article must bear the copyright notice and any header and footer information that IARIA applies to the published article.

IARIA grants royalty-free permission to the authors to disseminate the work, under the above provisions, for any academic, commercial, or industrial use. IARIA grants royalty-free permission to any individuals or institutions to make the article available electronically, online, or in print.

IARIA acknowledges that rights to any algorithm, process, procedure, apparatus, or articles of manufacture remain with the authors and their employers.

I, the undersigned, understand that IARIA will not be liable, in contract, tort (including, without limitation, negligence), pre-contract or other representations (other than fraudulent misrepresentations) or otherwise in connection with the publication of my work.

Exception to the above is made for work-for-hire performed while employed by the government. In that case, copyright to the material remains with the said government. The rightful owners (authors and government entity) grant unlimited and unrestricted permission to IARIA, IARIA's contractors, and IARIA's partners to further distribute the work.

A Survey of the Effectiveness of Automated Revenue Collection Systems in County Governments in Kenya

A Case Study of Kiambu and Taita Taveta County Governments

Margaret N. Njenga -Author
Project Coordinator,
@iLabAfrica, Strathmore University,
Nairobi, Kenya
mnjenga@strathmore.edu

Joseph Sevilla-Author
Director,
@iLabAfrica, Strathmore University,
Nairobi, Kenya
jsevilla@strathmore.edu

Abstract—**This paper examines the effectiveness of using an automated revenue collection system in two counties in Kenya: Taita Taveta and Kiambu County Governments. County governments in Kenya have recently adopted the automated revenue collection systems in order to aid in transparency and accountability of citizens' taxes. @iLabAfrica –Strathmore University provided the solution - CountyPro as a revenue collection system in both counties. The online system has a citizen portal and a government staff portal. This study focuses on the effectiveness of CountyPro system on the backend (staff) portal. A descriptive research design was adopted in order to provide answers to who, what, where, when and how. Questionnaires and interviews were used as the main tool for data collection. The strengths and weaknesses of both systems were also analyzed and also how they work together to bring out efficiency and effectiveness in the day to day operations of county governments. In general all county staff interviewed welcome the idea of an automated revenue collection system with Kiambu County recording a 60 percent increase in revenue as a result of using CountyPro. It is the perception of most county officials that the Point of Sale Terminal is the tool that is used to determine the revenue collected. However, this system by itself is not very effective. It has to be supported by an online system that will give a breakdown of all the kind of revenue that is collected. Counties should ensure that they have laid out proper infrastructure, trained its employees before rolling out of I.T projects. County management should also act as champions for change to help their employees transition from one system to another. Insights to this study will be used to improve on the current system and also help other counties in following the correct guidelines when coming up with a revenue collection system.**

Keywords— ***e-governance; revenue; management; effectiveness; system.***

I. INTRODUCTION

E-government is a generic term for web-based services from agencies of local, state and federal governments. In e-government, the government uses information technology and particularly the Internet to support government operations, engage citizens, and provide government services. The interaction may be in the form of obtaining information, filings, or making payments and a host of other activities via the World Wide Web [9]. Kenya's e-Government program was meant to address two impediments to development faced by many countries: endemic corruption and inefficiency [14].

Kenya has recently adopted a decentralized form of government. In the year 2010, Kenyans passed a new Constitution into place. One of the pillars of this new law is Devolution. Devolution refers to the transfer of decision-making capacity from higher levels in an organization to lower levels [4]. When governments devolve functions, they transfer authority for decision-making, finance, and management to quasi-autonomous units of local government with corporate status [2]. In Kenya, this meant that there was to be a national government and also the country was subdivided into forty seven counties. Each of these counties would have their own small governments and would each be headed by a governor. These counties generate their own revenue in addition to a percentage released by the national government. These small governments have also embraced incorporation of e-governance in their counties.

According to Heeks [6], e-government initiatives are important to governments in the following ways: they improve government processes or e-administration, help cut time and financial costs, manage employee and financial performance and create empowerment. They also help in connecting citizens to help improve the relationship between the government and its people. Citizens can hold public servants accountable for their decisions and actions. In turn public services are improved. Citizens' voices can also be heard and hence improved participation, and building external relations with members of the private sector through establishing meaningful partnerships.

As part of the devolution process, county governments are mandated to provide the essential services to its citizens such as good education, health care and good roads. Each county receives a percentage of funds from the national government depending on some factors such as poverty level index. This amount received from the central government is not sufficient to provide these services. To ensure sustainability, counties are expected to generate their own income. It is therefore very necessary to have a good and strong foundation of an automated revenue collection management system.

With this in mind, @iLabAfrica Strathmore University in partnership with Namu Health and iPay developed a software solution for County Revenue management. System requirements were discussed with Kiambu and Taita Taveta county governments and an agreement was reached to develop a citizen centric e-governance and revenue management solution. The system has a front end portal for the citizens and a backend portal for the county staff. The back end portal has various functionalities that include: permit and licenses processing, billing, property rates payment, enforcement functionality, ability to generate reports and grievances management. This study will focus on the effectiveness of the back end system to the county staff.

The reasons behind the development of this solution were to have a system that would enhance transparency and accountability of those in power. It is a tool that is used to aid those in government to use their instruments of power efficiently and effectively. Though it has several components to it, it works as a unified system. In addition to this, it is also expected to work optimally, have minimal errors, easy to use and seal loopholes that could have existed in the old Local Authorities Integrated Financial Operations Management System (LAIFOMS).

Below are the objectives of the research:

1. To determine if county employees understand the usefulness of a revenue collection system.

2. To determine the perception of county staff on County Pro Revenue Collection System.

3. To determine the strengths and weaknesses County Pro Automated Revenue Collection System.

4. To continually improve the revenue collection system as a whole based on the results of the research.

To examine the above requirements, a study was performed. The paper is structured as follows below. Section 2 outlines literature gathered on previous systems from books, journals and online research comparing automated revenue collection systems in other countries with CountyPro system deployed in Kenya. Section 3 describes the methodology used to collect and gather data and as well as how the sample size was reached and how the data was obtained. Section 4 interprets and discusses the significance of the findings of the research. Section 5 – Conclusion summarizes the findings of the research giving recommendations for future improvements.

II. LITERATURE REVIEW

Automation is a set of technologies that results in operation of machines and systems without significant human intervention and achieves performance superior to manual operation [1].

For government to match in performance with the growth and expectations of its constituents, it must dramatically increase its fiscal depth without incurring costly recurring overheads. Automated systems have been proven to be capable of introducing massive efficiencies to business processes that can result in increased revenue [8].

According to Sani [7], By automating revenue collection, service providers have better audit trail since all transactions captured can be detailed by time, whom and where. This prevents revenue loss through abuses as all moves are recorded electronically.

Problems such as high costs for collection, fraud, underpayment and leakages in revenue could be made worse by massively expanding the current taxable base without the use of adequate computerized solutions. The problems of tracking and identifying fraud or rogue revenue collectors are only compounded by the usage of manual or centralized systems due to the resources and overheads needed to monitor and control such problems. A decentralized, automated revenue collection system allows for increased and timely access to information that would otherwise take too much time and effort to generate from the available hard copy records [8].

Various scholars have analyzed several revenue collection systems. Gidisu [10] studied the automation system procedures of the Ghana customs division. A survey of 40 officials from the Customs Division with specific duties and responsibilities in automation system management at the Ghana Revenue Authority (GRA) was conducted. After this survey, it can be said that the automation is a powerful monitoring tool for GRA. It was realized that there was a positive impact of automation system usage and the cost of tax administration, automation and effectiveness of revenue collection.

According to Mitullah et al. [13], from a survey of 175 local authorities in Kenya, it was discovered that most of these local authorities had a lot of challenges in realizing their mandate for instance delivery of services. This was attributed to poor revenue management systems. The study concluded that information system was instrumental in enhancing and proper management of resources at the local authorities.

A study was conducted by Justus [3] to determine the effects of an integrated revenue collection system in Machakos County, as well as challenges facing its implementation. The study established that implementation of integrated revenue collection system influenced revenue collection positively. Challenges that were identified to influence implementation of integrated revenue collection system included resources, staff capacity, political interference and remoteness among others.

III. METHODOLOGY

A Descriptive Research Design approach was used for this study. This kind of study enabled the researcher provide answers to the questions of who, what, when, where, and how effective the automated revenue collection system has been in Kiambu and Taita Taveta counties. Quantitative and qualitative survey approaches were used where questionnaires and interviews were administered to county staff members. The respondents for this study included: billing officers, revenue officers, cashiers and parking attendants.

Both primary and secondary data sources were used to gather information. Literature was extracted from books, journals and online research by scholars.

County governments are further divided into sub counties, which are smaller divisions of the counties. Judgmental sampling was used in selecting two (2) sub counties from Kiambu County and two (2) sub counties from Taita Taveta County. This decision was reached by determining the sub counties in which there is the most activity while operating County Pro system.

Slovin's formula was used to determine the sample size from each of these counties.

$$n = \frac{N}{1 + Ne^2}$$

Where n=Sample size, N=Total population e= Desired Margin of error, that is 0.05% based on a 95% confidence level. The study included a total population of 150 staff members from Kiambu County. $n=150/(1+150(0.05)^2) = 109$. This generated a sample size of 109.

The same formula (Slovin's formula) was used to determine the sample size for staff of Taita Taveta County, where N = total population with a 95% confidence level. This generated $n=100/(1+100(0.05)^2) = 80$.

IV. DISCUSSION OF FINDINGS

The study sought to find out the effectiveness of a revenue collection system in both Kiambu and Taita Taveta Counties.

To perform this survey, 65% of the interviewees from Kiambu were male and 35% were female. In Taita Taveta County, 55% were male while 45% interviewees were female. In both counties, the male gender was more represented.

The study revealed that 2% of the respondents were below the age of 25 years, while 52% were between the age of 25 and 35 years. 32% were between ages 36 to 45 years. Only 10% of the respondents were above the age of 45. This shows that majority of the county staff are young, energetic and productive individuals.

Sixty seven (67%) percent of both the population selected in Taita Taveta and Kiambu Counties had been working in the county or the defunct local authority for more than ten (10) years while twenty eight (28%) percent had been working for between five and ten years. Thirty nine percent (39%) of the population had been working in the county government for less than five (5) years. This data shows the respondents' ability to compare the past and present systems and hence shows appropriateness to answer survey questions.

Fifty percent (50%) of the population from both counties agreed that they think about an online system when they hear about an automated revenue collection system, thirty percent (30%) think of a point of sale terminal while the rest of the population think of both. This is an indication that a significant number of the population still views the POS terminal as the main revenue collection system. The respondents further added that the reason behind this is that the POS was a tangible device. This made them feel as if they had shifted systems.

Below is a breakdown of some of the findings from the questionnaires issued to the county staff of Kiambu County.

TABLE I. KIAMBU COUNTY RESPONSES

Kiambu County					
	Strongly Disagree	**Disagree**	**Neutral**	**Agree**	**Strongly Agree**
System was easy to learn	0	0	10 (9.2 %)	19 (17.4%)	80 (73.3%)
System is Easy to use	0	0	4 (3.7 %)	30 (27.5%)	75 (68.8%)
Time based efficiency	0	7 (6.4%)	8 (7.3 %)	40 (36.7%)	54 (49.5%)
System has low downtime	0	0	1 (0.9 %)	6 (5.5%)	102 (93.6%)
Reports are easy to generate	0	8 (7.3%)	4 (3.7 %)	56 (51.4%)	41 (37.6%)
Reports are easy to understand	0	0	2 (1.8 %)	15 (13.8%)	92 (84.4%)
CountyPro serves better than old system	0	6 (5.5%)	15 (13.8 %)	70 (64.2%)	18 (16.5%)

Below is a breakdown of some of the findings from the questionnaires issued to the county staff of Taita Taveta County.

TABLE II. TAITA TAVETA COUNTY RESPONSES

Taita Taveta County					
	Strongly Disagree	**Disagree**	**Neutral**	**Agree**	**Strongly Agree**
System was easy to learn	0	1 (1.25%)	15 (18.8%)	4 (5%)	60 (75%)
System is Easy to use	0	2 (2.5%)	5 (6.25%)	20 (25%)	53 (66.25 %)
Time based efficiency	0	12 (15%)	5 (6.25%)	3 (3.75 %)	60 (75%)
System has low downtime	0	0	3 (3.75%)	6 (7.5%)	71 (88.75 %)
Reports are easy to generate	0	0	4 (5%)	40 (50%)	36 (45%)
Reports are easy to understand	0	0	5 (6.25%)	20 (25%)	55 (68.75 %)
CountyPro serves better than old system	0	5 (6.25%)	11 (13.75 %)	30 (37.5 %)	34 (42.5 %)

73.3% from Kiambu County and 75% of the respondents from Taita Taveta County Strongly agreed that the system was easy to learn. These strong percentages from users from both counties are an indicator that the system satisfies the simplification of the system to basic users.

A notable 49.5% of respondents from Kiambu County and 75% of respondents from Taita Taveta County strongly agreed that the system is efficient in terms of helping them save time. This was done in comparison to the older system.

93.6% of Kiambu County employees indicated that the system experienced low downtime while 88.75% of the employees from Taita Taveta County indicated that the system had experienced low downtime while operating it. They also added that the only times when there was a lag with the system was during the deadline for renewal of the yearly permits. This is during the 31st day of March of every year.

16.5% of respondents from Kiambu County strongly agreed that reports generated from CountyPro system are easy to understand while 68.75% of the respondents from Taita Taveta County agreed that the reports are easy to understand. The reason for this difference was attributed to the little amount of training that had been conducted to the employees of Kiambu County at that time as compared to Taita Taveta County. Also, most of the trainees introduced to these new systems were not technologically savvy.

Eighty percent (80%) of the population from both counties agreed that the CountyPro online system has facilitated an increase in revenues.

When asked about effect of the system on corruption, the respondents stated that the system had sealed some loopholes in terms of county employees not handling money physically. They further stated that the use of mobile money, VISA and MasterCard options helped to seal some of these loopholes.

On the other hand, Taita Taveta County interviewees stated that, ethics, good leadership and enforcement is what had contributed to the reduction.

80 percent agreed that the new system served them better than the old system – LAIFOMS system while 20 percent thought that it did not serve them better than the previous system.

The researcher further sought to find out if the previous LAIFOMS system had any drawbacks. It was pointed out that it had several major challenges. These included:

1. LAIFOMS was a standalone system that only catered for the backend operations, leaving citizens with inaccessibility problems, with nowhere to get information on county rates and charges, the finance act etc. This made them constantly visit the county offices which took a lot of time.
2. Money could not be tracked to the LAIFOMS system as bank slips were not posted to the system. Instead the banking slips were recorded manually in a register outside the system.
3. Collection of rates and money was done using physical receipts hence making the systems vulnerable to corrupt practices by county officials.

The researcher also sought to find out the advantages of the new automated system by the county employees.

One major advantage of this system was the ability of county employees to access data from anywhere as the system was online. With the system, citizens could access information on procedures and services with the availability of documents online such as the Public Finance Act. It also incorporated various modes of payment as opposed to only cash option which was in the old system. With the new automated system, citizens could apply for permits and download them online once they are processed by county officials on the back end system. Below are some of the other benefits of CountyPro that were pointed out by the county officials:

1. It helped to stamp out bureaucracy and endless procedures as in the old system.
2. The system helped reduce paper work.
3. Reports in the system could be accessed at any time.
4. The reports showed best and worst performances in revenue and hence knew where to put in more resources. This helped in decision making by the policy makers.
5. Anytime anywhere access to the Government, 24 hours a day 365 days a year. Even those living in the diaspora could make payments for fees and licences owed online.
6. Efficient delivery of government services to citizens, and reduced dependence on human interactions.
7. Reduced cost in terms of time and resources in processing transactions/applications and delivering citizen services.
8. Increased citizen participation/empowerment through transparency, and access to information.
9. Citizens felt a form of participation in the county government as the system sent them alerts whenever their permit was ready for collection at the county offices. It also alerted them whenever there were any waivers or balances in revenues.
10. Upsurge in revenues.

The respondents were asked whether they thought the system had brought about an increase in revenue in their counties and why. They stated that the increase in revenue was facilitated in the following ways:

1. Monitoring and management of cash flow. All the sources of revenue are centrally located and are therefore easy to monitor.
2. Helped to recover lost revenues. After using the system, the county officials could more easily identify the customers who owed the county and were able to claim it.
3. Citizens could process their payments themselves without having to be followed up hence increased revenues.
4. The ability to monitor revenue trends through specific and detailed reports. It became possible to make projections and come up with attainable targets. This in turn helped boost the morale of the employees.

When asked what they would have liked to see different with the system, the respondents suggested that the training time for the system be increased to help them fully understand and internalize the system.

The respondents further added that there was need to have civic education, in order to educate citizens on the new way of doing things. This is because they had faced some resistance during the introduction of the system, as a result of introduction of new fees and charges.

Intensive ICT education needed to be conducted on the county employees in order for them to adapt as fast as possible. Most of the employees were 30 years old and above and were not ICT savvy.

The POS gadgets were being acquired from different suppliers and some of them would turn out faulty. This resulted into lags in revenue collection and overworking of county employees.

The most common problems reported with the new system is inadequate Internet connection and poor network connectivity. Another challenge that was encountered is the lack of automation of all modules in the system. For example, considering health is a devolved unit, the health module in County Pro system should be integrated with all County Hospitals. This is to ensure that all the revenue that all the revenue is collected through a single system. The respondents further suggested that this kind of integration would lead to more transparency and accountability.

V. CONCLUSION AND FUTURE WORK

Most of the respondents' years of service was between five to 10 years and hence were familiar with most of the operations at the county offices and the different kinds of revenue collected. The whole population interviewed shows that the new automated system was accepted by the county officials. Most of the respondents recommended County Pro system for its robustness, ease of learning, friendliness, effectiveness, ability to make work easier and facilitating an increase in revenues.

An increase in revenue in both Kiambu and Taita Taveta counties was realized out of the adoption of the automated revenue collection systems. A report released by Commission on Revenue Allocation (CRA) that reviewed counties indicated that Kiambu was among top five devolved units that made great improvements in revenue collection. Strathmore University carried out extensive research in the county on how we have collected revenue, gave us a raft of recommendations before deploying our new County Pro System that is helping seal loop holes. Kiambu County government has praised its partnership with Strathmore University on revenue collection automation, saying the deal had contributed to success of financial reforms being witnessed in the county [12]. County employees including county management have the perception that an automated system for revenue collection is a point of sale terminal. This therefore leads to too much concentration on only the Point of Sale gadget hence the flopping of the online system. County management should focus on both the point of sale gadget as well as the online system as both will in the end work together to achieve the success of the project.

County governments should work to ensure that there are the necessary resources and infrastructure before the rolling out of automation projects. Internet connectivity should also be ensured for smooth flow of work. A lot of training for the county officials should be conducted before the roll out of such projects. County governments should also ensure that all employees who will participate in an automation project are Tech Savvy. This will reduce training time, and in turn further reduce support costs.

In order to realize maximum benefits and revenue from the system, county management should look towards automating all modules. For example, the Agriculture, Livestock and Fisheries module, Liquor licence and health modules.

The transition to the new automated system was a step by step process. The county government management as well as the employees needed to identify other revenue streams that were not captured in the system to ensure that they are also automated.

In as much as the system was functioning optimally, there was still need to catch up with technology. This would involve the speeding up of cashless solutions in all the county regions such as payment via mobile money. This would solve the problem of failure of some POS gadgets.

REFERENCES

[1] IIT. Kharagpur, *Power Electronic.* [Online], 2nd Ed., India: Aldipdg. [Retrieved: November, 2015], from: http://nptel.ac.in/courses/Webcourse-contents/IIT%20Kharagpur/Industrial%20Automation%20control/pdf/L-01(SM)(IA&C)%20((EE)NPTEL).pdf

[2] J. Litvack, What is Decentralization. [Retrieved: August, 2016] from http://www.ciesin.org/decentralization/English/General/Different_forms.html

[3] M. Justus, Effects of an Integrated Revenue Collection System and Challenges Facing Its Implementation in Machakos County, Unpublished MBA Thesis, University of Nairobi, Kenya, 2014.

[4] MAB/MIAC, *Devolution of Corporate Services*, No. 6, April 1992, pp. 3.

[5] R. Babbie, The Practice of Social Research. 12th ed. Belmont, CA: Wadsworth Cengage, 2010.

[6] R. Heeks, *eGovernment for Development. Success and Failure in eGovernment Projects.* [Retrieved: August, 2015], from http://www.egov4dev.org/success/definitions.shtml

[7] R. Sani, Automating revenue collection New Straits Times, Mar 30, 2009(pp. 49– 88), in S. Cnossen & H-W. Sinn (eds) Public Finance and Public Policy in the New Century (Cambridge, Mass.: MIT Press), 2009.

[8] S. Guido, Community Revenue Collection System. Journal of Community Informatics, vol 4, no. 3, pp. 2-3, 2008.

[9] S. Sharma, Assessing E-government Implementations, Electronic Government Journal, vol 1 no 2, 2004, pp. 198-212, 2004.

[10] T. Gidisu, Automation System Procedure of the Ghana Revenue Authority on the Effectiveness of Revenue Collection: A Case Study of Customs Division, Unpublished MBA Thesis, Kwame Nkrumah University of Science and Technology, 2012.

[11] University of Southern California, Organizing your Social Sciences Research Paper. Types of Research Designs. [Retrieved November, 2015] from http://libguides.usc.edu/writingguide/researchdesigns

[12] J. Wanzala, Kiambu County lauds Strathmore's IT Solution. Standard Digital. [Retrieved: August, 2015] from http://www.standardmedia.co.ke/business/article/2000173952/kiambu-county-lauds-strathmore-s-it-solution?articleID=2000173952&story_title=kiambu-county-lauds-strathmore-s-it-solution&pageNo=1

[13] W. Mitullah, MO. Akivaga, *Management of resources by local authorities*: The case of Local Authority Transfer Fund. Nairobi: CLARIPRESS, 2005.

[14] W. Mwololo, T, and M. Winnie, Kenya E-Government. [Retrieved: November, 2015], from https://newmediadev2009.wikischolars.columbia.edu/Kenya E-Government, 2009

Scoring of Machine-Learning Algorithms for Providing User Guidance in Special Purpose Machines

Valentin Plenk*, Sascha Lang†, Florian Wogenstein‡
Institute of Information Systems at Hof University, Hof, Germany
Email: *valentin.plenk@iisys.de, †sascha.lang@iisys.de, ‡florian.wogenstein@iisys.de

Abstract—**We propose a system to make complex production machines more user-friendly by giving the operator recommendations, such as "in the last 10 occurrences of this event the operators performed the following keystrokes". We describe algorithms to generate the recommendations based on data on former user-interaction and process values and to store them in a knowledge base. We also propose algorithms to retrieve recommendations suited to the current process state. We evaluate their performance on simulated data and data gathered from real production machines.**

Keywords–machine-learning; human machine interfaces; special-purpose machines; production machines

I. Introduction

In [1] we proposed to apply machine learning algorithms to generate recommendations like "in the last 10 occurrences of this event the operators performed the following keystrokes" for operators of complex production machines. Figure 1 shows the context: our system, the black box, builds a knowledge base from past operator interactions with a complex production machine.

The system faces two challenges: One is to retrieve the recommendations that are most suitable for the current state of the production machine. The other is to build the knowledge base by extracting the operator interactions from data logged during production.

Despite of the idea being quite straightforward there is apparently little similar work. Most of the work in the production-machine sector applies machine learning to applications dealing with pattern detection in process data or distingushing different datasets [2]–[5]. The algorithms are used to flag errors or problems with production quality to the machine's operator who needs expertise to counter the detected problem.

Challiol et al. [6] describes an application striving to display content dependent on the users context. While addressing a completely different application domain this is similar to our approach in terms of matching context to content. The authors assume the content to be available and do not propose specific context matching algorithms. The context matching algorithm proposed in [7] is quite abstract as is its performance evaluation presented in [8].

Our approach as described in Section II covers aspects similar to [6], [7] and [8] with strong focus on the application in complex production machines. Furthermore we propose to automatically generate the content of the recommendations.

The algorithms presented in Section III build a knowledge base that can be edited. i.e., the recommendations can be presented to an experienced operator who can modify or delete them. This is a marked difference to most machine learning algorithms whose knowledge base cannot be edited.

Section IV evaluates the performance of these algorithms on test data. These data are used to train the algorithms and to derive the "truth" we compare to the generated recommendations.

Section VI summarizes the results and gives a short outlook for our future work and the challenges expected in the near future.

II. Proposed Approach

The basic idea of our approach is to extract operator knowledge from the continuous stream of PLC-variables logged during the operation of the machine. We want our algorithms to be as generic as possible and to work without deeper understanding of the machine. However, we need information to recommend actions to the machine operator: Which columns

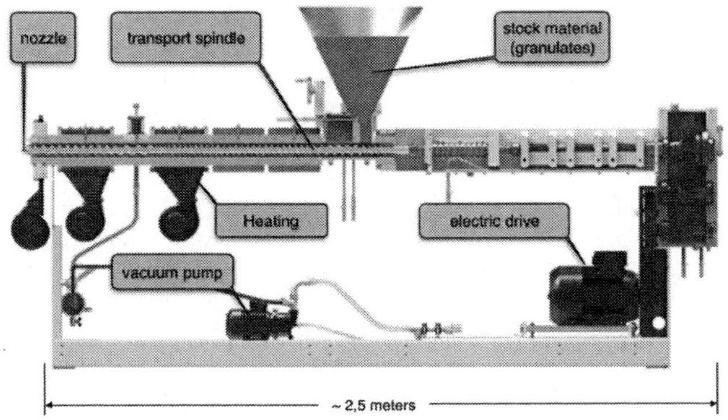

a) Plastics extruder

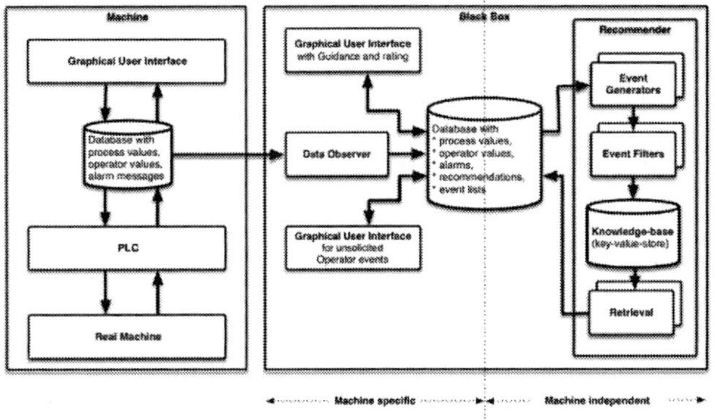

b) System architecture

Figure 1. Application context for our machine learning algorithms

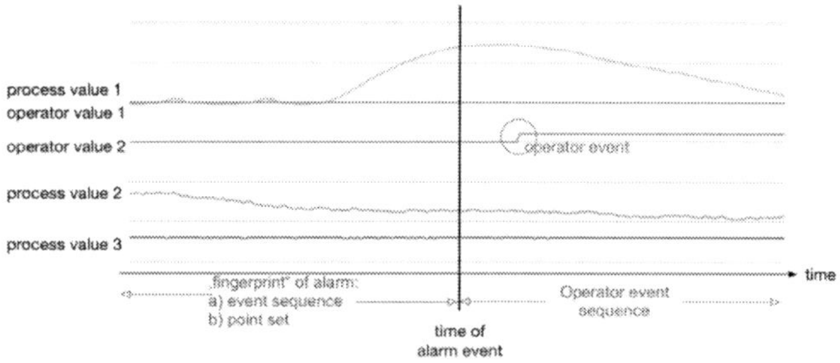

Figure 2. Principle of user guidance generation

in the PLC variable database correspond to values affected by the process, e.g., actual temperature, and which are controlled by the operator, e.g., temperature set-point. We call the first set of variables process values and the second set operator values. The rows of the database hold consecutive values for these variables.

With this information we propose to generate user guidance as shown in Figure 2: The alarm messages in the alarm database are used to tag discrete parts of the endless stream of data logged in the PLC-variable database. The alarm tag and the data before the occurrence of the alarm are regarded as a fingerprint that identifies the alarm. This fingerprint consists of process and operator variables. The data after the occurrence of the alarm also consists of process and operator values. The actions performed by the operator will change some of these operator values. By detecting these changes we build a sequence of operator events. This sequence represents the actions of the operator corresponding to a fingerprint.

With fingerprints and corresponding operator sequences we build a key value store representing the operators' expert knowledge. The fingerprint is the key and the operator sequence the value. These key-value pairs are generated by monitoring the continuous stream of data logged during the operation of the machine. When the machine raises an alarm the algorithms generate a key for that new problem. Now the key-value pairs in the knowledge base are searched for the best matching elements. The best matches are then provided as a recommendation on how to solve the problem. The fingerprint and the operator sequence performed by the operator are then stored in the knowledge base as a new key value pair.

In Section III we describe two generally different approaches for the processing of the fingerprints: One set of algorithms directly uses the process and operator variables. The other set converts the data in the fingerprint into an event sequence. Events are generated for the parts of a time series where the data changes. Several events for one or several columns constitute an event sequence. Detailed descriptions of the event generation can be found in [1] and [9].

III. Recommendation Generation

In case the machine needs operator assistance, i.e., it raised an alarm by adding a new row in the error message table, our guidance generation algorithm is triggered. The fingerprint of the current situation is used as a search key for the knowledge base.

Section III-A describes a set of algorithms that converts the data in the fingerprint into an event sequence and then searches for this sequence. Section III-B introduces the best performing algorithm that directly uses the N-dimensional point set of process and operator variables in the fingerprint as a key. More algorithms are described in [9]

A. Recommendation Generation using Event Sequences

1) Map: In [1] we used a content addressable memory, i.e., a Java map, to store the knowledge base. We build the knowledge base by iterating through all past occurrences of alarms. For each fingerprint we calculate the event sequence and use it as the key and the corresponding operator event sequence as the value. Each key-value pair is then stored in the map. For ambiguous entries we store the frequency of the respective sequence in the past.

To generate a recommendation we simply generate the event sequence corresponding to the fingerprint of the current alarm and query the map for this key. For map entries with several values for one key we return several recommendations and their frequency in the past. Should the key not be in the map, we cannot generate a recommendation.

2) Map with Statistical Event Filtering: The map presented in Section III-A1 will only find a recommendation if the search key exactly matches one of the stored keys. If the event sequences contain spurious events, e.g., caused by noise, we can not find a match. So we created an algorithm to suppress the irrelevant events.

We use a statistical approach identifying unimportant events to remove them from the process event sequence. The filtered event list is then processed as described in Section III-A1.

The filtering is done in two steps. First we iterate through the event sequences of all stored fingerprints for one alarm and count how often an event is contained in the set of event sequences. In the next step we remove all events with a frequency below a defined threshold from the event sequences. So far our tests indicate that 0.5 is a reasonable choice.

3) String matching: The map algorithm presented in Section III-A1 requires the key in the query to be absolutely identical to one key in the map. Thus, it is very sensitive to changes in the key, i.e., the event-sequence. We alleviate this rigorous selection criterion by storing all keys and the corresponding values in a vector. We then loop through all the elements and compare the sequence of the query to the sequence stored in the vector. We return the value as a recommendation that is most similar to the key in the query.

We evaluate the similarity with the Levenshtein distance for string values as described in [10]. For the comparison we treat each event in the sequence as one letter.

B. Recommendation using point sets

The approach described in Section III-A3 allows for disturbance by not requiring a complete match between search key and stored key. We can take this idea one step further by regarding the data points in the fingerprint as a set of n-dimensional points. For one timestamp every process and operator value is one dimension.

As in Section III-A3 we store the point sets and the corresponding operator events in a vector of key-value-pairs. To generate a recommendation we loop through all the elements of this vector and compare the fingerprint, i.e., the point set, of the query to the point sets stored in the vector. We return the value, i.e., the operator events, as a recommendation that is most similar to the key in the query.

Plenk et al. [9] details several procedures for matching point sets. In this paper we only detail the procedure that performed best on our data. It is a combination of the Hausdorff distance for point sets and the Manhattan-distance for points.

The Hausdorff-distance defines a similarity between two not empty point sets A and B. For one point a that is an element of set A the distance between a and B is defined as:

$$D_H(a, B) = \min_{b \in B} d(a, b) \tag{1}$$

where d is the Manhattan-distance between two points a and b with the dimension i:

$$d_m(a, b) = \sum_i |a_i - b_i| \tag{2}$$

On that basis the the Hausdorff-distance between two sets A and B can be put down as:

$$d_h(A, B) = \max\left\{\max_{a \in A} D(a, B), \max_{b \in B} D(b, A)\right\} \tag{3}$$

IV. Quality of Recommendations

To evaluate our algorithms we use a "batch-mode" that basically iterates through a long time series of PLC-variables and alarms. It triggers the recommender for each alarm encountered during the iterations and requests a recommendation. This recommendation is compared to the operator sequence stored in the time series. This sequence is taken as the expected recommendation or "truth" for this particular alarm. If the recommendation is equal to the operator sequence in the test-set the test is positive and C_{pos} is incremented. If a different sequence is returned or the algorithm returns nothing the test fails and C_{neg} is incremented. With these two parameters we can calculate the Quality Q_{alg} for the algorithm:

$$Q_{alg} = \frac{C_{pos}}{C_{neg} + C_{pos}} \cdot 100\% \tag{4}$$

We need to train the recommender algorithms before feeding them alarms. We generate the training data by dividing the dataset into four subsets with an equal number of alarms. One of these sets is taken as the test-set. The others are combined into a training set. Thus, we can run four tests on each of our algorithms by using the test-sets as training data. The remaining sets of alarms and data are used to train the algorithms.

The expression in equation 4 ranges between $0 \leq Q_{alg} \leq 100\%$. A closer look at the test-sets however reveals cases where an operator sequence only occurs in the test set and not in the training set. Thus, the algorithm cannot learn this recommendation and is not able to detect it correctly. Some other keys have more than one associated value. In these cases, the algorithm is not able to decide which operator sequence is the right one. Consequently the maximum possible score is less than 100%.

For our evaluation we use the data gathered during 5 minutes before the alarm to generate the fingerprint. The operator sequence is generated for the duration of the alarm, i.e., for the period of time in which the alarm condition is true.

TABLE I. Dataset 1: The five most frequently raised alarms

Alarm number	Alarm count	With op-sequence	Different op-sequences
1101	171	57	12
1012	107	86	19
12	106	53	19
22	106	49	16
2	59	30	10

A. Dataset 1 – Simulation

As a first dataset we use a slightly extended version of the data we used in [1]. This dataset was obtained by operating a simulation model of the production machine. It contains $N_{Rows} = 1,150,063$ rows and $N_{Alarms} = 1,254$ alarms.

As in [1] we focus on alarm 1101. There are 171 instances of this alarm in the dataset. For 57 of these instances we could generate the 12 different operator sequences shown in Table II. The remaining 114 occurrences of the alarm do not contain operator sequences because the simulation was not always operated properly.

8 of these operator sequences corresponding to 10 occurrences of alarm 1101 are only contained in one of the four subsets. Thus, these sets are either part of the training data or of the test data. We therefore reduce the maximum possible score from 57 to 47.

For the event based algorithms we found 3 ambiguous event sequences. Table VII shows all event sequences. The ambiguous event sequences have more than one corresponding operator sequence. We consider ambiguous sequences as unlearnable. Our algorithm will always recommend the sequence with the highest frequency. Therefore we subtract the number of the other sequences from the maximum possible. In our case this reduces the maximum possible by 10 to 37.

For event sequence based algorithms we get a maximum possible score of

$$Q_{alg_{1ev_{max}}} = \frac{37}{57} \cdot 100\% \approx 65\% \tag{5}$$

For the point based algorithms we get a maximum score of

$$Q_{alg_{1p_{max}}} = \frac{47}{57} \cdot 100\% \approx 82\% \tag{6}$$

B. Dataset 2 – Converted Data from real Machine

The second set of process- and operator values was taken from a production machine.

The database of that machine does not yet conform to our interface standard and thus did not log the alarms. We resorted

TABLE II. Dataset 1: Operator sequences

Frequency	Operator Sequence	Learnable
36	Loop1_3_SP_170	yes
7	Loop1_2_SP_170#Loop1_3_SP_170	yes
4	Loop1_4_SP_170	yes
2	Loop1_3_SP_250	no
1	Loop1_3_SP_225	no
1	AOUT1_1_OutValue_10	no
1	AOUT1_1_OutValue_29	no
1	AOUT1_1_OutValue_50	no
1	AOUT1_2_OutValue_9	no
1	Loop1_7_SP_250	no
1	Loop1_2_SP_250#Loop1_3_SP_250	no
1	Loop1_3_SP_170#Loop1_4_SP_170	no

to generating our own alarms based on process experts' definitions for lower and upper bounds of process values. The first occurrence of a value being outside the bounds was the starting time for an alarm. The stopping time was taken the moment the value was back inside the bounds.

The resulting database contains $N_{Rows} = 1,223,992$ rows describing the machine state and $N_{Alarms} = 5,667$ alarms. Table III shows the five most frequent alarms in the dataset. With the machine logging the data every 10 seconds we have a runtime of $\approx 3,400$ hours. This means we have 1.7 alarms per hour. According to our project partner this number seems to be very high.

Not all alarms we generated from the machine were useful for our test. Some alarms were not followed by operator events. We assume that these alarms are artifacts of our data preparation algorithm. Other alarms obviously occured at the end of a production shift and consequently lead to the recommendation to shut down the machine. We chose alarm 225 for our evaluation.

Alarm 225 occurred 1949 times. For 277 instances of this alarm we could create a fingerprint and an operator sequence.

We assume that this difference is partly due to our self generated alarms and partly due to spurious and short alarms that disappear without user intervention.

For these 277 instances of alarm 225 we generated the 22 operator sequences shown in Table IV. 13 of these operator sequences corresponding to 46 event sequences occur in only one of the four subsets. Consequently we reduce the theoretical maximum by 46. Furthermore, we found 33 ambiguous event sequences. We subtract all but one of these ambiguous sequences except for those that are already marked as not learnable because they appear in only one subset. So we have to subtract further 47 alarms.

So for event based algorithms we get a maximum score of:

$$Q_{alg_{2ev_{max}}} = \frac{184}{277} \cdot 100\% \approx 66\% \quad (7)$$

For the point based algorithms we get a maximum score of:

$$Q_{alg_{2p_{max}}} = \frac{231}{277} \cdot 100\% \approx 83\% \quad (8)$$

C. Dataset 3 – Real Data

The third data set was gathered on a new production machine equipped with our system. The machine being new it is not yet fully productive and there is no long history of operation. We logged $N_{Rows} = 417,663$. As in the previous dataset a new row of data has been logged every 10 seconds. So we got a total run time of $\approx 1,200$ hours. During this time period we found $N_{Alarms} = 2,277$ alarms.

TABLE III. DATASET 2: THE FIVE MOST FREQUENTLY RAISED ALARMS

Alarm number	Alarm count	With op-sequence	Different op-equences
225	1949	277	22
423	1059	70	43
122	763	243	93
123	503	149	69
214	238	142	81
213	109	76	60

TABLE IV. DATASET 2: OPERATOR SEQUENCES FOR ALARM 225

Frequency	Operator Sequence	Learnable
54	AOut1_1_85#	yes
53	AOut1_1_84#	yes
30	AOut1_1_83#	yes
30	AOut1_1_86#	yes
20	AOut1_1_81#	yes
17	AOut1_1_82#	yes
16	AOut1_1_77#	no
14	AOut1_1_87#	yes
11	AOut1_1_80#	yes
9	AOut1_1_76#	no
7	AOut1_1_78#	no
5	AOut1_1_79#	no
2	AOut1_1_88#	yes
1	AOut1_1_0#AOut2_1_6#	no
1	AOut1_1_83#AOut2_1_64#	no
1	AOut1_1_83#AOut2_1_98#	no
1	AOut1_1_84#AOut2_1_98#	no
1	AOut1_1_85#AOut2_1_98#	no
1	AOut1_1_87#AOut2_1_73#	no
1	AOut1_1_89#	no
1	AOut1_1_91#Loop2_2_SP_230#	no
1	AOut2_1_99#	no

For this particular machine we can determine if the machine is in the production process or is idling. So we filtered for alarms being raised while the machine was running, we found $N_{Alarms} = 1,393$ alarms during operation. For 412 of these we could determine an operator sequence. Along with that we found 381 alarms which only occurred for a short time period, in this case less than 10 seconds. Table V shows the five most frequent alarms.

For further investigation we chose alarm 21.5 which occured 321 times. For 45 instances of the alarm we were able to create a fingerprint and an operator sequence. We used these alarms for our test. Table VI shows the operator sequences we could determine. In total we found 31 different operator sequences.

These operator sequences cover 17 alarms that we consider as learnable. The other ones occurring only once are not learnable by our algorithms. From the unique operator sequences 28 occurred in only one set. 28 alarms belong to these operator sequences. In this dataset we did not find any ambiguous event sequences. So for event and point based algorithms we get a maximum score of:

$$Q_{alg_{3ev_{max}}} = Q_{alg_{3p_{max}}} = \frac{17}{45} \cdot 100\% \approx 38\% \quad (9)$$

V. COMPARISON OF THE ALGORITHMS

After processing the three datasets through all of our algorithms we calculated the score for each dataset according to eq. 4. As discussed in sec. IV the resulting values for Q_{alg} are not comparable between the datasets because of unlearnable and ambiguous operator event sequences.

TABLE V. DATASET 3: THE FIVE MOST FREQUENTLY RAISED ALARMS

Alarm number	Alarm count	With op-sequence	Different op-equences
21.5	321	45	31
42.0	121	9	9
36.0	106	14	14
38.0	82	2	2
18.2	46	24	22
17.3	43	28	25

TABLE VI. DATASET 3: OPERATOR SEQUENCES FOR ALARM 21.5

Frequency	Operator Sequence	Learnable
13	ButtonDrive_1	yes
2	ButtonDrive_1# OutScrewspeed_30	yes
2	ButtonDrive_1# ToolButtonTemp_1	yes
1	ButtonDrive_0	no
1	ButtonDrive_0# MeltpumpOut_30	no
1	ButtonDrive_1# ButtonTemp_1# MeltpumpOut_20	no
1	ButtonDrive_1# ButtonTemp_1# MeltpumpOut_30# ToolButtonTemp_1	no
1	Button.Drive_1# OutScrewspeed_11	no
1	Button.Drive_1# OutScrewspeed_13# MeltpumpOut_10	no
1	Button.Drive_1# OutScrewspeed_14# MeltpumpOut_50	no
1	Button.Drive_1# OutScrewspeed_15# MeltpumpOut_13	no
1	Button.Drive_1# OutScrewspeed_19# MeltpumpOut_11# ToolTemp3SP_180	no
1	Button.Drive_1# OutScrewspeed_24# MeltpumpOut_37	no
1	Button.Drive_1# OutScrewspeed_27# ToolTemp1SP_195# ToolTemp2SP_185# ToolTemp3SP_180	no
1	Button.Drive_1# OutScrewspeed_28# Temp2SP_220#Temp4SP_220# ToolButtonTemp_1# ToolTemp1SP_195# ToolTemp3SP_198	no
1	Button.Drive_1# OutScrewspeed_43# MeltpumpOut_17	no
1	Button.Drive_1# OutScrewspeed_48	no
1	Button.Drive_1# OutScrewspeed_51	no
1	Button.Drive_1# OutScrewspeed_59# Meltpump.Out_40	no
1	Button.Drive_1# HaulOffButton_0	no
1	Button.Drive_1# MeltpumpOut_0	no
1	Button.Drive_1# MeltpumpOut_12	no
1	Button.Drive_1# MeltpumpOut_24# ToolButtonTemp_1	no
1	Button.Drive_1# MeltpumpOut_30	no
1	OutScrewspeed_15	no
1	OutScrewspeed_32	no
1	OutScrewspeed_34	no
1	Meltpump.Out_0	no
1	Meltpump.Out_14	no
1	Meltpump.Out_26	no
1	Meltpump.Out_48	no

Eqn. 5 to 9 give the maximum possible score for each dataset. With this score we can normalize the results as

$$Q_{alg_{rel}} = \frac{Q_{alg}}{Q_{alg_{max}}} \cdot 100\% \qquad (10)$$

To put the performance into perspective we calculate the performance of a simple approach that always recommends the most frequent user sequence in the knowledge base. Such an approach would propose a correct operator sequence for all the alarms associated with the most frequent operator sequence:

$$Q_{simple} = \frac{100\%}{N_{OpSeq}} \cdot N_{mostFreqOpSeq} \qquad (11)$$

i.e.,

$$Q_{simple_1} = \frac{100\%}{57} \cdot 36 \approx 63\% \qquad (12)$$

$$Q_{simple_2} = \frac{100\%}{277} \cdot 54 \approx 19\% \qquad (13)$$

$$Q_{simple_3} = \frac{100\%}{45} \cdot 13 \approx 28\% \qquad (14)$$

Figure 3 shows the normalized performance of the different algorithms on the three datasets and for reference the normalized scores of the simple approach.

The mapping algorithm introduced in [1] and III-A1 scores $\approx 63\%$ on dataset 1. On dataset 2 it performed less with only $\approx 12\%$. On dataset 3 it did not give any correct recommendations. We attribute this lack of performance to the fact that this dataset does not have two identical event sequences. Since the algorithm performs a one to one matching it is not capable

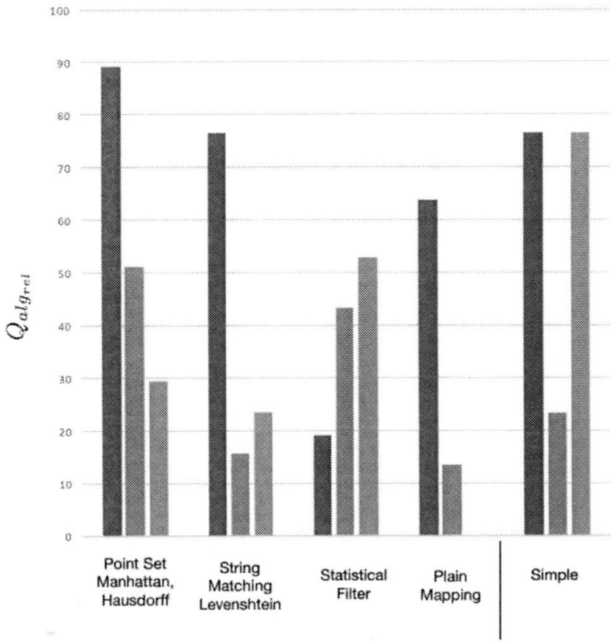

Figure 3. Normalized results for various recommender algorithms (left bars: Dataset 1, middle bars: Dataset 2, right bars: Dataset 3)

of finding any solution. In total it always scores less than the simple approach.

With $\approx 42\%$ on dataset 2 the statistical filter performs better than the simple approach and is almost on par with the point based algorithms. It also shows the best performance on dataset 3 but is still behind the simple algorithm.

The point based algorithm scores better on dataset 1 with almost 90% and $> 50\%$ on dataset 2. Dataset 3 however proves difficult with $\approx 29\%$.

VI. CONCLUSION AND FUTURE WORK

We used three datasets to generate and evaluate recommendations in "batch-mode". On some datasets our recommender algorithms outperformed a simple algorithm by a factor of 2. On our newest and smallest dataset, i.e., dataset 3, they lacked performance. That being said, we want to point out that all algorithms need the operator sequences we generate from the logged data.

Nevertheless, we will need to look further into dataset 3 and enlarge our knowledge base from $\approx 1,200$ hours and $\approx 2,000$ alarms to at least 5,000 hours and 10,000 alarms.

A quick analysis of dataset 3 showed that many alarms occur at the same time or in a very short timespan. So we will investigate whether it is easier to find a recommendation for groups of alarms.

We started interviewing experienced operators. We learned that in some situations the appropriate operator action is to just wait for the alarm to clear itself. So it might be a good idea to include a "do nothing" recommendation in the knowledge base.

We also plan to give the operator the possibility to evaluate our recommendation and to factor that evaluation into our recommendation.

References

[1] V. Plenk, "Improving special purpose machine user-interfaces by machine-learning algorithms," in Proceedings of CENTRIC 2016 : The Ninth International Conference on Advances in Human-oriented and Personalized Mechanisms, Technologies, and Services, Rome, August 2016, pp. 24 – 28.

[2] P. K. Wonga, J. Zhonga, Z. Yanga, and C. M. Vong, "Sparse bayesian extreme learning committee machine for engine simultaneous fault diagnosis," in Neurocomputing, 2016, pp. 331 – 343.

[3] K. Zidek, A. Hosovsky, and J. Dubjak, "Diagnostics of surface errors by embedded vision system and its classification by machine learning algorithms," in Key Engineering Materials, 2015, pp. 459 – 466.

[4] J. Luo, C.-M. Vong, and P.-K. Wong, "Sparse bayesian extreme learning machine for multi-classification," in IIEEE Transactions on Neural Networks and Learning Systems, 2014, pp. 836 – 843.

[5] A. Ziaja, I. Antoniadou, T. Barszcz, W. J. Staszewski, and K. Worden, "Fault detection in rolling element bearings using wavelet-based variance analysis and novelty detection," Journal of Vibration and Control, vol. 22, no. 2, February 2016, pp. 396–411.

[6] C. Challiol, A. Fortier, S. Gordillo, and G. Rossi, "A Flexible Architecture for Context-Aware Physical Hypermedia," in 18th International Workshop on: Database and Expert Systems Applications, 2007. DEXA '07., September 2007.

[7] S. Sigg, S. Haseloff, and K. David, "Context Prediction by Alignment Methods," in Proceedings of the 4th International Conference on Mobile Systems, Applications, and Services (MobiSys 2006), June 2006.

[8] S. Sigg, D. Gordon, G. von Zengen, M. Beigl, S. Haseloff, and K. David, "Investigation of Context Prediction Accuracy for Different Context Abstraction Levels," IEEE Transactions on Mobile Computing, vol. 11, no. 6, June 2012, pp. 1047 – 1059.

[9] V. Plenk, S. Lang, and F. Wogenstein, "Providing user guidance in special purpose machines by machine-learning algorithms," To appear in International Journal On Advances in Software, vol. 10, no. 3 and 4, December 2017.

[10] M. P. J. van der Loo, "The stringdist Package for Approximate String Matching," The R Journal, vol. Vol. 6/1, 2014, pp. 111 – 122.

[11] J. Lunze, Ereignisdiskrete Systeme. München: Oldenbourg, 2006.

TABLE VII. DATASET 1: EVENT SEQUENCES AND OPERATOR SEQUENCES

EventSequence	Operator Sequence	Frequency	Learnable
Empty[1]	Loop1_3_SP_225	1	no
	Loop1_3_SP_250	2	no
	AOUT1_1_OutValue_29	1	no
	Loop1_3_SP_170	2	yes
	Loop1_2_SP_170#Loop1_3_SP_170	1	yes
	Loop1_2_SP_250#Loop1_3_SP_250	1	no
	AOUT1_1_OutValue_10	1	no
Loop1_4_PV_B#Loop1_4_PV_X	Loop1_4_SP_170	1	yes
Loop1_3_SP_0.0	Loop1_3_SP_170	28	yes
	Loop1_2_SP_170#Loop1_3_SP_170	6	yes
Loop1_4_PV_H#Loop1_4_SP_50.0# Loop1_4_PV_G#Loop1_4_PV_F	Loop1_4_SP_170	1	yes
Loop1_4_SP_0.0	Loop1_4_SP_170	2	yes
Loop1_2_SP_0.0#Loop1_4_SP_0.0	Loop1_3_SP_170#Loop1_4_SP_170	1	no
	Loop1_3_SP_170	2	yes
Loop1_3_SP_10.0	Loop1_3_SP_170	1	yes
Loop1_4_PV_H#Loop1_5_PV_H# Loop1_6_PV_H#Loop1_7_SP_50.0	Loop1_7_SP_250	1	no
Loop1_3_SP_60.0	Loop1_3_SP_170	3	yes
AIN2_1_Value_G#MP1_Value_G#MT1_Value_F# MP1_Value_F#AIN2_1_Value_F#MP1_Value_B# MT1_Value_E#MP1_Value_C	AOUT1_2_OutValue_9	1	no
MP1_Value_E#MP1_BandMin_E#MP1_BandMax_E# Loop1_6_PV_E#Loop1_6_PV_F#Loop1_6_PV_E# Loop1_6_PV_D#Loop1_6_PV_C#Loop1_6_PV_B# Loop1_6_PV_A#MP1_BandMax_F#MP1_Value_H# MP1_BandMax_H#MP1_BandMin_F	AOUT1_1_OutValue_50	1	no

[1] In our simulation empty event sequences can occur when an alarm is generated by changing the alarm condition. This happens during testing or demonstration.

TABLE VIII. DATASET 2: AMBIGIOUS EVENT SEQUENCES

Event-Sequence	Operator-Sequence	Frequency	Learnable
1	AOut1_1_84#	2	yes
	AOut1_1_82#	1	yes
2	AOut1_1_86#	3	yes
	AOut1_1_87#	1	yes
	AOut1_1_85#	1	yes
3	AOut1_1_86#	2	yes
	AOut1_1_87#	1	yes
4	AOut1_1_84#	5	yes
	AOut1_1_85#	4	yes
5	AOut1_1_83#	2	yes
	AOut1_1_85#	2	yes
	AOut1_1_84#	1	yes
6	AOut1_1_80#	2	yes
	AOut1_1_79#	1	no
7	AOut1_1_87#	1	yes
	AOut1_1_86#	1	yes
8	AOut1_1_85#	1	yes
	AOut1_1_84#	1	yes
9	AOut1_1_83#	1	yes
	AOut1_1_84#	1	yes
10	AOut1_1_84#	2	yes
	AOut1_1_83#	1	yes
11	AOut1_1_82#	1	yes
	AOut1_1_83#	2	yes
12	AOut1_1_86#	2	yes
	AOut1_1_88#	1	yes
	AOut1_1_87#	1	yes
13	AOut1_1_87#	1	yes
	AOut1_1_86#	1	yes
14	AOut1_1_86#	2	yes
	AOut1_1_87#	1	yes
15	AOut1_1_85#	2	yes
	AOut1_1_86#	2	yes
16	AOut1_1_84#	1	yes
	AOut1_1_83#	1	yes
	AOut1_1_85#	1	yes
	AOut1_1_82#	1	yes
17	AOut1_1_83#	1	yes
	AOut1_1_85#	7	yes
	AOut1_1_84#	3	yes

Event-Sequence	Operator-Sequence	Frequency	Learnable
18	AOut1_1_86#	2	yes
	AOut1_1_87#	1	yes
19	AOut1_1_85#	5	yes
	AOut1_1_85#AOut2_1_98#	1	no
	AOut1_1_84#	1	yes
20	AOut1_1_82#	1	yes
	AOut1_1_81#	1	yes
21	AOut1_1_85#	1	yes
	AOut1_1_87#	1	yes
22	AOut1_1_87#	3	yes
	AOut1_1_85#	1	yes
	AOut1_1_86#	1	yes
23	AOut1_1_83#	1	yes
	AOut1_1_81#	1	yes
24	AOut1_1_81#	2	yes
	AOut1_1_80#	1	yes
25	AOut1_1_85#	1	yes
	AOut1_1_86#	1	yes
26	AOut1_1_80#	3	yes
	AOut1_1_81#	1	yes
27	AOut1_1_84#	1	yes
	AOut1_1_85#	1	yes
28	AOut1_1_77#	3	no
	AOut1_1_78#	1	no
29	AOut1_1_85#	1	yes
	AOut1_1_86#	1	yes
	AOut1_1_84#	2	yes
30	AOut1_1_82#	1	yes
	AOut1_1_81#	1	yes
31	AOut1_1_85#	1	yes
	AOut1_1_86#	1	yes
32	AOut1_1_83#	1	yes
	AOut1_1_85#	1	yes
33	AOut1_1_84#	1	yes
	AOut1_1_83#	1	yes
	AOut1_1_85#	2	yes

Person Re-identification in Crowded Scenes with Deep Learning

Yupeng Wang, Huiyuan Fu, Shuangqun Li
Beijing Key Lab of Intelligent Telecommunication Software and Multimedia
Beijing University of Posts and Telecommunications
Beijing, China
E-mail: {wyp, fhy, lsq}@bupt.edu.cn

***Abstract*—Person re-identification in crowded scenes is very important. Most images come from different surveillance video and cameras, and one person may look different in a variety of scenes, viewpoints, lighting and so on. The existing methods have limited effects in practical applications. In this paper, we propose a convolutional neural network for person re-identification in crowded scenes. The model structure of this network combines pedestrian detection and re-identification. In addition, we propose a loss function to better match the target person by calculating Pearson correlation evaluation. The experimental results show that our method is effective.**

Keywords-deep learning; person re-identification; crowded scenes; convolutional neural network; loss function

I. Introduction

Person re-identification means to match the target person in different images. These images come from different cameras, so the challenges lie in the different backgrounds, the changes of human postures, camera viewpoints, lighting, occlusions and so on. Person re-identification can be applied to surveillance video, such as cross-camera searching and tracking of criminals [1], which is very helpful for public safety. Therefore, person re-identification draws the attention of scholars and a lot of researches have been done in recent years [3][7-10][12].

Person re-identification consists of two steps: pedestrian detection and re-identification. Many researches have been done in the field of pedestrian detection. Dalal et al. [4] proposed Histogram of Oriented Gradient (HOG) feature for human detection. It can describe the edge features of human body. Zhu et al. [5] extracted Multi-scale Intrinsic Motion Structure features for pedestrian detection. These methods use hand-crafted features and linear classifiers to detect persons. Hand-crafted features may lose some important information of the original images, and the classification results are not good enough. In recent years, Deep Learning has attracted much attention in the fields of image, audio, natural language processing and so on. Yang et al. [6] proposed Convolutional Channel Features which achieved good performances in pedestrian detection. Zhang et al. [2] used Region Proposal Network (RPN) followed by boosted forests on high-resolution convolutional feature maps. As for person re-identification, the common method is part matching. Zhao et al. [7] adopted patch matching and estimated patch salience. Zheng et al. [10] proposed a PoseBox structure, which pose is estimated by affine transformations. However, these methods may introduce errors in part detection. Yi et al. [8] and Varior et al. [9] used the siamese convolutional neural network for person re-identification. Most of the existing person re-identification methods assume perfect pedestrian detections. In fact, these hand-cropped bounding boxes are unavailable in the applications of realistic scenes. Xiao et al. [3] proposed an end-to-end framework for person re-identification. Also, Yamaguchi et al. [13] used a natural language query to handle this task. But these methods are still difficult to satisfy complex realistic scenes. In these methods, the scenes are simple and contain only 1 to 5 people. The effect is not good in the scenes with many people.

In order to deal with person re-identification in crowded scenes, we propose a convolutional neural network (CNN) which combines pedestrian detection and re-identification. It can learn more effective deep features due to CNN with deeper network layers. In addition, we propose a loss function to better match the target person. The similarity of the target person and persons in crowded scene images is calculated by Pearson correlation evaluation. Finally, with fine-tuning on the weights initialization, experimental results on Large Scale Person Search dataset (PSDB) [3] show that the proposed method gains new state-of-the-art performances.

II. Person Re-identification in Crowded Scenes

Our work consists of two parts: deep learning network model and matching loss function for person re-identification. In the detection phase, we construct residual network units (ResNet) [11] and RPN [14]. In order to converge better and faster, we fine-tune the network via Xavier [16] filler instead of Gaussian filler. And in the re-identification phase, we construct aggregated residual network units (ResNeXt) [15] to extract the deep features of detected persons. After that, we propose a loss function to match the target person by calculating the similarity between the target person features and detected person features.

A. Model Structure for Person Re-identification

The model structure for person re-identification in crowded scenes is shown in Fig. 1. We construct ResNet-50 [11] as our base CNN model to extract image features. The kernel size is 7 in the first convolutional layer with 64 channels. We perform Batch Normalization (BN) layer after convolutional layers and Rectified Linear Units (ReLU) layer is performed after each BN layer. After the layer, there are three blocks. In the first block, there are three residual units which include three convolutional layers with 1, 3, and 1 of kernel size respectively. In the second block, there are four residual units. In the third block, there are three residual units. These convolutional layers are different in channels. The residual units can produce convolutional feature maps with 1024 channels.

Next, we need to get the person bounding boxes. An RPN [17] with a Softmax classifier is added to get 9 anchors and predict whether each bounding box is a person or not. It selects the top 128 bounding boxes as final proposals of persons in the image.

The weights initialization method of the RPN is Gaussian filler. Weights are randomly drawn from Gaussian distributions with fixed mean (e.g., 0) and fixed standard deviation (e.g., 0.01). This is the most common initialization method in deep learning. To make the information in the network flow better, the variance of each layer's output should be equal. Xavier [16] filler makes weights a uniform distribution with the mean of 0 and the variance of *Var*, as follows

$$Var = \frac{N}{n_i + n_{i+1}}, \tag{1}$$

where n_i is the i-th input number, and n_{i+1} is the i+1-th input number, as well as the i-th output number. By default, the variance takes into account only the number of inputs ($n_{i+1} = 0$, $N = 1$). But the number of inputs and outputs is often unequal, and the variance takes into account only the number of outputs ($n_i = 0$, $N = 1$). For balanced consideration, $N = 2$. In our network, Xavier filler is adopted to make the effect better.

Then, the task is to match the target person in these proposals. An RoI-Pooling layer is added for each proposal. And this layer links two blocks of ResNeXt-50(32×4d) [15]. In the first block, there are three aggregated residual units, which the second convolutional layer is grouped convolutions with 32 groups. In the second block, there are three aggregated residual units which are different from units of the first block in kernel output channels. Finally, we add the matching layer to calculate the similarity of these features with the target person by our proposed loss function.

B. Matching Loss Function for Person Re-identification

To match the target person in an image, we store the features of all people in this image, and calculate the similarity of these features with the target person. If one similarity is the largest, then the corresponding person is likely to be the target. In the labeled boxes, the features of the target person box is denoted as x, where $x \in R^D$ and D is the feature dimension, that is, x is a feature vector of D dimensions. The feature of one person box in the image is denoted as y^l, where $y^l \in R^D$ and D is the feature dimension. During the forward propagation, we compute the Pearson correlation coefficient C_j^l between the target features and the box features with the j-th class ($j \in [1, L]$ and L is the number of the boxes), as follows

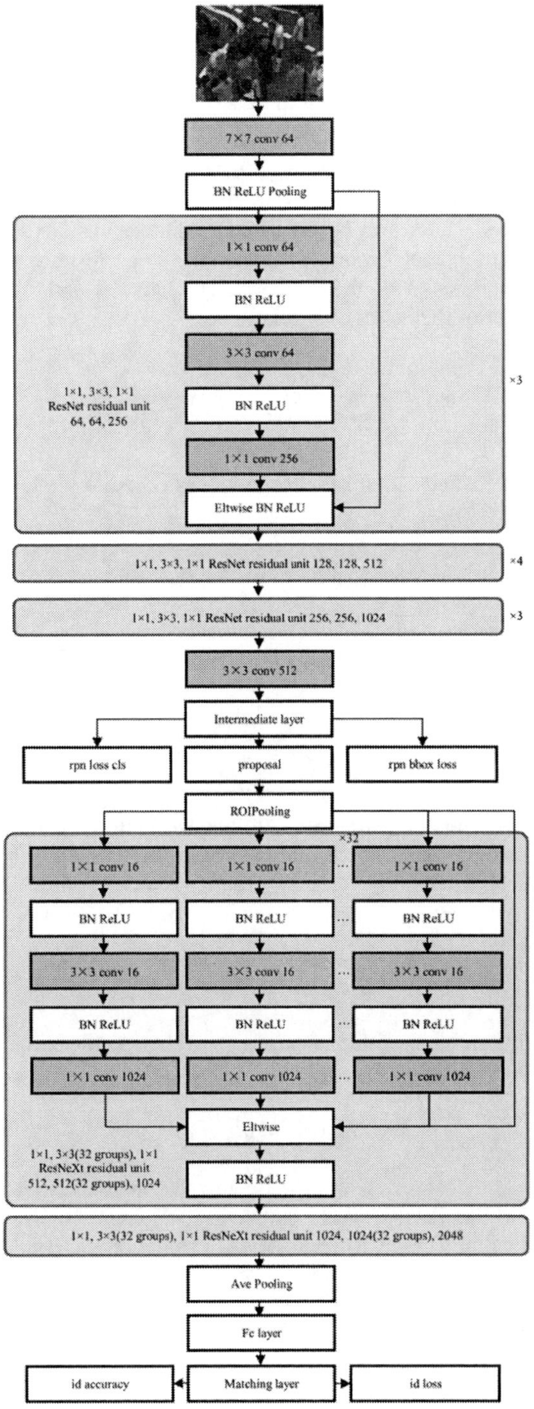

Figure 1. Our model structure.

$$C_j^l = \frac{\sum_{i=1}^{D}(x_i-\bar{x})\left(y_i^l-\overline{y^l}\right)}{\sqrt{\sum_{i=1}^{D}(x_i-\bar{x})^2}\sqrt{\sum_{i=1}^{D}\left(y_i^l-\overline{y^l}\right)^2}}, \tag{2}$$

where $\bar{x}$ and $\bar{y^l}$ are the mean values of x and y^l. During the backward propagation, we update y^l_j by

$$y_j^l \leftarrow \gamma y_j^l + (1-\gamma)x\,, \tag{3}$$

where $\gamma \in [0,\ 1]$. Similarly, in the unlabeled boxes, the feature of one person box in the image is denoted as y^u, where $y^u \in R^D$ and D is the feature dimension. The Pearson correlation coefficient C^u_j between the target features and the box features with the j-th class ($j \in [1,\ U]$ and U is the number of the boxes), as follows

$$C_j^u = \frac{\sum_{i=1}^{D}(x_i-\bar{x})\left(y_i^u-\bar{y^u}\right)}{\sqrt{\sum_{i=1}^{D}(x_i-\bar{x})^2}\sqrt{\sum_{i=1}^{D}\left(y_i^u-\bar{y^u}\right)^2}}\,, \tag{4}$$

where $\bar{x}$ and $\bar{y^u}$ are the mean values of x and y^u. The y^u_j is updated by

$$y_j^u \leftarrow \gamma y_j^u + (1-\gamma)x\,. \tag{5}$$

So, the probability of the feature vector x as the i-th class labeled person is

$$p_i^l = \frac{e^{(C_i^l/\tau)}}{\sum_{j=1}^{L} e^{(C_j^l/\tau)} + \sum_{k=1}^{U} e^{(C_k^u/\tau)}}\,, \tag{6}$$

where τ makes softer probability distribution. In the same way, the probability of the feature vector x as the i-th class unlabeled person is

$$p_i^u = \frac{e^{(C_i^u/\tau)}}{\sum_{j=1}^{L} e^{(C_j^l/\tau)} + \sum_{k=1}^{U} e^{(C_k^u/\tau)}}\,. \tag{7}$$

In summary, the Pearson correlation matching loss function is defined as follows:

$$\mathrm{L} = \mathrm{E}(\log p_t^l)\,, \tag{8}$$

where $t \in [1,\ L]$. The Pearson correlation coefficient is the cosine similarity with subtracting average. It is more applicable to data using different evaluation criteria. Then, the loss effectively compares the features of the target person with all the features of the boxes and finds the person whose feature vector is the most similar with the target one from the crowded scene images.

III. EXPERIMENTS

A. Dataset

We use PSDB [3], a large-scale dataset for person re-identification which covering hundreds of scenes. There are 18,184 images, 8,432 identities, and 96,143 pedestrian bounding boxes in this dataset. The training set contains 11,206 images and 5,532 query persons. The test set contains 6,978 images and 2,900 query persons.

TABLE I. PERSON RE-IDENTIFICATION METHODS

Methods	Top-1(%)	mAP(%)
DSIFT[17]+Euclidean	39.4	34.5
DSIFT[17]+KISSME[19]	53.6	47.8
LOMO+XQDA[18]	74.1	68.9
OIM[3]	78.5	75.4
Ours	**79.8**	**76.7**

TABLE II. WEIGHTS INITIALIZATION METHODS

Methods	Top-1(%)	Top-5(%)	Top-10(%)	mAP(%)
OIM (Gaussian)	78.5	90.1	92.3	75.4
OIM (Xavier)	79.0	90.0	92.8	76.0
Ours(Gaussian)	79.3	90.2	92.9	76.1
Ours(Xavier)	**79.8**	**90.3**	**92.9**	**76.7**

B. Experimental Results

For comparisons, we use some feature representations and the state-of-art method (Online Instance Matching, OIM) [3] for person re-identification on PSDB. The results are summarized in Table 1. Dense scale-invariant feature transform (DSIFT) [17] method is based on unsupervised salience learning. Euclidean distance metric and KISSME (keep it simple and straightforward metric) [19] distance metric are used for DSIFT. And Cross-view Quadratic Discriminant Analysis (XQDA) distance metric is used for Local Maximal Occurrence (LOMO) [18]. We use two kinds of evaluation metrics: cumulative matching characteristics (CMC top-K) and mean averaged precision (mAP). CMC top-K is the probability that the result hits within K times. And mAP is obtained by the average for precision results of each class of people.

The results are summarized in Table 1. Our method outperforms the others. On the one hand, the detector is also important for person re-identification. Most existing methods focus only on cropped images, so they may not be suitable for re-identification in the crowded scene images. On the other hand, different distance metric methods also have an impact on the experimental results.

In addition, we compare the different weights initialization methods with OIM [3]. As shown in Table 2, Xavier filler is superior to Gaussian filler in most cases. These results are relevant to specific network models and tasks.

The samples of our method in the experiment are shown in Fig. 2. The inputs are an image of the target person and some images with people. The outputs are the result images of bounding boxes with re-identification similarities. The person with maximum similarity is the target. Both in non-

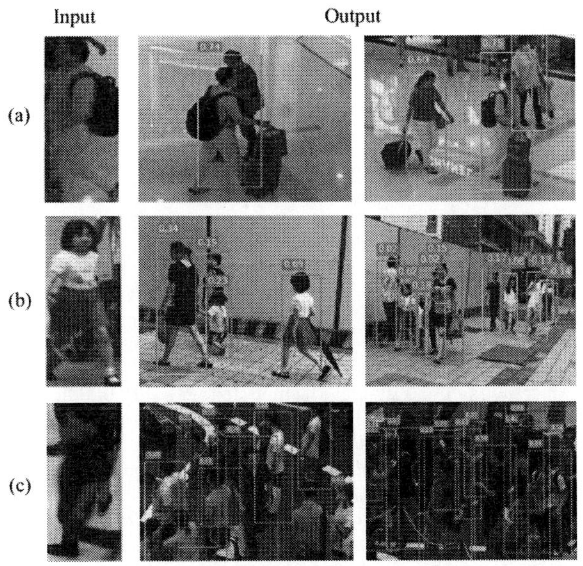

Figure 2. Samples of our method for person re-identification. (a) Non-crowded, (b) mid-crowded, (c) crowded scenes.

crowded and crowded scenes, our method can well identify the target person with the largest similarity.

IV. CONCLUSION AND FUTURE WORK

In this paper, we propose a deep learning framework for end-to-end person re-identification. We adopt one CNN to link the two tasks of pedestrian detection and person re-identification. And ResNeXt residual units are constructed in the model to extract features more effectively. We also propose a Pearson Correlation Matching loss function to match the target person. Compared with existing methods, the performance of our method is improved with fine-tuning through experiments on PSDB dataset.

In the future, we plan to study person re-identification for surveillance video. The application of video is necessary for industry and society. On the one hand, many images make up video frame sequences. That is, the methods of images can be extended to video. On the other hand, there are new and available features due to the continuity of video frames. We will also combine person re-identification with Natural Language Processing to achieve a wider range of applications in the future.

ACKNOWLEDGMENT

The research reported in this paper is supported by the National Natural Science Foundation of China under Grant No.61402048; The NSFC-Guangdong Joint Found under No.U1501254.

REFERENCES

[1] X. Wang, "Intelligent multi-camera video surveillance: A review," Pattern Recognition Letters, vol. 34, no. 1. pp. 3–19, 2013.

[2] L. Zhang, L. Lin, X. Liang, and K. He, "Is faster R-CNN doing well for pedestrian detection?," in European Conference on Computer Vision. Springer International Publishing, vol. 9906 LNCS, pp. 443–457, 2016.

[3] T. Xiao, S. Li, B. Wang, L. Lin, and X. Wang, "Joint Detection and Identification Feature Learning for Person Search," ArXiv e-prints, 2017.

[4] N. Dalal and B. Triggs, "Histograms of oriented gradients for human detection," in Proceedings - 2005 IEEE Computer Society Conference on Computer Vision and Pattern Recognition, vol. I, pp. 886–893, 2005.

[5] J. Zhu, et al., "Pedestrian detection in low-resolution imagery by learning Multi-scale Intrinsic Motion Structures (MIMS)," in Proceedings of the IEEE Computer Society Conference on Computer Vision and Pattern Recognition, pp. 3510–3517, 2014.

[6] B. Yang, J. Yan, Z. Lei, and S. Z. Li, "Convolutional channel features," in Proceedings of the IEEE International Conference on Computer Vision, vol. 11–18–Dece, pp. 82–90, 2015.

[7] R. Zhao, W. Ouyang, and X. Wang, "Person re-identification by salience matching," in Proceedings of the IEEE International Conference on Computer Vision, pp. 2528–2535, 2013.

[8] D. Yi, Z. Lei, S. Liao, and S. Z. Li, "Deep metric learning for person re-identification," in Proceedings - International Conference on Pattern Recognition, pp. 34–39, 2014.

[9] R. R. Varior, M. Haloi, and G. Wang, "Gated Siamese Convolutional Neural Network Architecture for Human Re-Identification," ECCV16, pp. 1–18, 2016.

[10] L. Zheng, Y. Huang, H. Lu, and Y. Yang, "Pose Invariant Embedding for Deep Person Re-identification," ArXiv e-prints, 2017.

[11] K. He, X. Zhang, S. Ren, and J. Sun, "Deep Residual Learning for Image Recognition," in 2016 IEEE Conference on Computer Vision and Pattern Recognition (CVPR), pp. 770–778, 2016.

[12] W. S. Zheng, et al., "Partial person re-identification," in Proceedings of the IEEE International Conference on Computer Vision, vol. 11–18–Dece, pp. 4678–4686, 2015.

[13] M. Yamaguchi, K. Saito, Y. Ushiku, and T. Harada, "Spatio-temporal Person Retrieval via Natural Language Queries," ArXiv e-prints, 2017.

[14] S. Ren, K. He, R. Girshick, and J. Sun, "Faster R-CNN: Towards Real-Time Object Detection with Region Proposal Networks," IEEE Trans. Pattern Anal. Mach. Intell., vol. 39, no. 6, pp. 1137–1149, Jun. 2016.

[15] S. Xie, R. Girshick, P. Dollar, Z. Tu, and K. He, "Aggregated Residual Transformations for Deep Neural Networks," CoRR, vol. abs/1611.0, 2016.

[16] X. Glorot and Y. Bengio, "Understanding the difficulty of training deep feedforward neural networks," Proc. 13th Int. Conf. Artif. Intell. Stat., vol. 9, pp. 249–256, 2010.

[17] R. Zhao, W. Ouyang, and X. Wang, "Unsupervised salience learning for person re-identification," in Proceedings of the IEEE Computer Society Conference on Computer Vision and Pattern Recognition, pp. 3586–3593, 2013.

[18] S. Liao, Y. Hu, X. Zhu, and S. Z. Li, "Person re-identification by Local Maximal Occurrence representation and metric learning," in Proceedings of the IEEE Computer Society Conference on Computer Vision and Pattern Recognition, pp. 2197–2206, 2015.

[19] M. Kostinger, M. Hirzer, P. Wohlhart, P. M. Roth, and H. Bischof, "Large scale metric learning from equivalence constraints," in Proceedings of the IEEE Computer Society Conference on Computer Vision and Pattern Recognition, pp. 2288–2295, 2012.

Dynamic Scrutable User Modeling utilizing Machine Learning

Dima S. Mahmoud
ADAPT Centre
School of Computer Science and Statistics, Trinity College Dublin
Dublin, Ireland
dima.mahmod@adaptcentre.ie

Abstract— **Personalization generally attempts to offer information and services that are delivered to meet user's individual preferences. It helps by providing the appropriate services in a dynamic and automatic manner. Involving the user in this process may enhance how tailored information and services are delivered. However, there is a challenge in engaging users in the user modeling process. Building user models in a manner that engages the user in a feedback cycle may improve the quality of the model and the user's control over the personalization. Allowing such user control over machine learning-derived user models is a significant research challenge as such models are often difficult to scrutinize. This is the main challenge addressed in this early stage research. This work proposes using ontology-based domain models to provide a means for users to engage with ML-derived models. Moreover, such an approach may enable the user model scope to grow as a user's preferences grow.**

Keywords-Personalization; User Modeling; Machine Learning; Scrutability.

I. INTRODUCTION

Personalization has the potential to play an important role in supporting the tailoring of system behavior in ways that fit each individual user's preferences. Personalization reduces information overload and facilitating targeted access to relevant information objects in an information system [1]. However, it depends on creating and maintaining an adequate set of information about the user's preferences [2].

Generally, personalization is achieved whenever the behavior of a system is adjusted by information about the user [3]. Postma and Brokke defined personalization as "a segmented form of communication that sends (groups of) different recipients different messages tailored to their individual preferences" [4] and this is the definition adopted in this work. A key challenge lies in delivering the right information at any time, at any place, while respecting the user interests [5]. These interests are continuously evolving and changing, generally at a rate that is faster than most implicit approaches, which create a user model from observed behavior in a system, can keep pace with.

Capturing the right information about the user at the right time in the right way is the main concept of user modeling [5][6]. A user model is the collection of personal information associated with a specific user. So, it is the fundamental basis for any personalization changes to a system. It is also potentially a key instrument through which the user can adjust and control how a personalization system works for them.

Machine learning has recently attracted much attention and has been employed for user modeling [7]. Much research is concerned with utilizing machine learning for building intelligent user models [8]. Generally, the internet evolution was the motivating force underlying the recent surge of research in this field [9]. However, the user is not engaged in most of these studies leading to models which are difficult to represent to a user and nearly impossible for users to control. M. E. Muller summarized the idea in one sentence: "machine learning in user modeling tries to mimic a user's behavior, but it does not model a user" [8]. We believe that modeling the user, while providing them with control over the model, is the main contribution in the work proposed in this paper.

The research outlined in this paper is concerned with developing an approach that incorporates dynamic machine learning with user scrutability to facilitate effective user modeling. The ML processes will operate over a large corpus of email messages that will enable the modeling of a variety of dimensions of the users in the corpus. There are three main aspects under consideration in this study: user preferences, actionable items and machine learning techniques. This combination is responsible for building the user model, taking into consideration the user's control over the user model.

This paper is structured as follows: Section II discusses the background and related work. Section III states the research question and the intended contribution. After that, the design of the proposed solution is outlined and the final section will state the progress of this work so far.

II. BACKGROUND AND RELATED WORK

This section discusses personalization and user modeling in more detail. It then illustrates the relationship between them.

A. *Personalization and User Modeling*

A fundamental objective of research in personalization and adaptation is to make systems more usable, more useful, and to provide users with experiences fitting their specific background knowledge and objectives [10]. What is really challenging in a world loaded with a large volume of information, is not only to make information available at anytime, anywhere and in any form, but to precisely specify the "right" thing, at the "right" time and in the "right" way.

Personalization has become of major concern across all industries and interest in tailoring customer/user experiences has increased significantly [11]. In general, it is hard to have a broader definition of personalization [3]. It is mainly the concept of using a user profile that may include personal information and sometimes their preferences. Accordingly, this has received significant consideration both as a way for offering appealing services and locking these users into the appropriate services [12].

Once information is collected about a certain user, the system can evaluate that data by a preset analytical algorithm and then personalize it to meet the user's needs [13]. This adaptation considers every feature of the system's behavior. The user profile affects how information and functions are displayed. In the case of [13], profiles highlight only relevant aspects and hide knowledge that is not needed by the user; this is in addition to providing offers and proposals [14].

There are three main mechanisms for handling system personalization. To date, personalization and profile modeling have mainly fallen into two categories: explicit and implicit techniques [15]. However, the hybrid approach was added and followed afterward [2][16].

- Explicit modeling: In the beginning, the user was the main item in the personalization process. It depends completely on them to set their personal information and manage their interests. Explicit personalization may reduce somehow the effort of resource management, as well as assuring the data precision However, managing an entire preference set manually means putting a burden on the user to carry out the profile management responsibilities [17]. In other words, the user has the mission to update their profile whenever a new service is encountered in order to refine and update interests in their profile.
  This approach fundamentally engages the user in the whole task and places the onus on the user. They set their profile manually in order to keep their interests up to date (maybe through an appropriate GUI). Although in this technique, the user is in charge of the whole management process, the responsibility of maintaining such potentially large profile is a burden. This can often lead to a sparse preference set and hence inaccurate personalization [2]. In fact, this is the main shortcoming of this mechanism as this undermines the strength of personalization.
- Implicit modeling: This approach is considered the other extreme in the personalization process. It primarily uses various techniques for monitoring and learning the user's preferences without engaging the user directly. The system tends to maintain the user profile and the preferences set on behalf of the user. This depends mainly on the intelligence level of the system. This may affect the information accuracy and accordingly the environment personalization [18].
- Hybrid implicit and explicit modeling: The learning approaches employed by such systems often fall under two types: rule-based learning algorithms (which store preferences as rules) and network algorithms (which store preferences in some network structure). Rule-based learning algorithms have the advantage that their output is easily translated into a human-readable form allowing their knowledge to be understood [16].

The benefit of the hybrid approach is the minimal burden on the user, however, care must be taken to provide some method of user control. Without such functionality, the user cannot alter system behavior to reflect new situations or behaviors in a rapid way. Therefore, for more successful systems, hybrid personalization providing implicit personalization must be employed where possible, but also providing a GUI through which the user can manually manipulate their preferences and take final control.

B. User Models

Capturing the right thing at the right time in the right way is the main concept of user modeling [5][6]. A user model is the collection of personal information associated with a specific user. So, it is the basis for any personalized changes made by a system.

This research is concerned mainly with saying the "right" thing. Selecting which data is used and employed in the model depends on the goal of the application and the output required. It can include personal information [19] such as users' names, ages, their interests, their skills and knowledge, their goals and plans, their preferences and their dislikes, or data about their behavior and their interactions with the system. There are different design patterns for user models, though often a mixture of them is used.

- Static user models: Static models are the primary type of user models. As the name indicates, the model does not change; once data is collected they are normally not changed again. Changes in user's information do not affect the model and no learning algorithms are used to alter the model [5][20].
- Dynamic user models: Dynamic models allow a more active representation of users' preferences. The model can dynamically adapt to the different shifts in users' interests or their interactions with the model. This adaptation helps meeting the different needs of the users [5][20][21].
- Static and Dynamic (SaD) User Models: We can easily imagine that SaD as a hybrid modeling technique. It can be more common to allow the modeler to move from just using a one level standard static user model that uses static unchangeable information to a more thorough two-level model. This combines a static with a dynamic user model thus containing user preferences alters and various interactions with the system [20].

Since user modeling is fundamentally based on personal information, user control over the model may bring value, particularly when pertinent information about the user may not be implicitly observed by the system. Such a model with user control mechanisms is described as scrutable model [22]. It is designed so that a person has the option, but not an obligation, to determine what is modeled about them and how it is used.

Personalization generally aims to unburden the user of profile and preferences management tasks, adjusting the system to meet the user interests, thus leaning more towards implicit modeling with no user control. Also, in order to enhance the interaction between the user and a personalized environment, information systems must be capable of dynamically personalizing the system content [23]. For this reason, many pervasive and ubiquitous systems include machine learning techniques coupled with user behavior monitoring systems to provide the implicit personalization part, where preferences can be created and managed which may be coupled with user control (explicit engagement) for better results [1][2].

III. RESEARCH QUESTION AND INTENDED MAIN CONTRIBUTIONS

The ultimate goal of this research is building a dynamic scrutable user model while utilizing machine learning. The key idea of this work is to merge the three aspects: dynamic user modeling, scrutable user modeling, and machine learning. We believe that this combination would provide the optimum balance between implicit modelling, dynamic model growth towards new domains of interest and user control. The target is defining an approach to facilitate this balance.

The strategy of this study will start by building a case study as shown in Figure 1 - which is discussed in detail later - that demonstrates the idea as a proof of concept. Then the next step is the evaluation of the approach and presenting the results and comparing the results with and without the user control and discussing how far the user control over the user model affects the results.

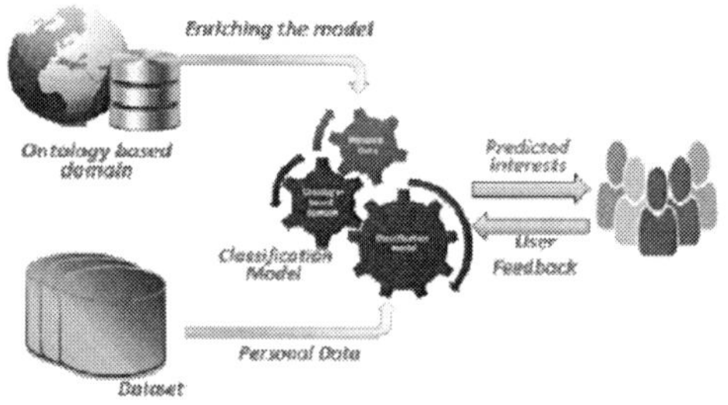

Figure 1. Classification model design

The case study has been designed to provide and promote to users services that suit their interests with taking in consideration the user feedback. We are working on this target by exploring large set of user data [24]. This data contains important and unimportant information for these participants. This work needs to process a very large number of facts and capture the vital chunks from it. This selected important data indicates each user's likes and dislikes. This information then goes through the learning phase in order to build a customized user model. The model will be capable of selecting the appropriate service for each user. These promotions should suit each user preferences.

The second aspect that plays the main role in this research is the system re-learning. In other words, how is user control exerted and maintained over the model? How to blend the machine learning and user control is envisaged a key outcome of this work.

The other dimension in this research is what these services are and how we can feed the model with different alternatives for the same subject. We are proposing to use an ontology-based domain, which uses terms from a domain model to indicate a user's relationship to different concepts. The Domain Model describes how concepts are connected to each other defining a semantic relationship between them [25][26]. We believe that using this ontology-based approach will enrich the system with various substitution potentials. This is discussed in detail later in the design section.

The main data source for this case study was selected to enable user profiles to be built from users' email messages. This is used to construct the initial user models. These emails are rich with personal information: inbox messages, sent messages, folders messages, outgoing messages, email subject, and time stamps are all used. The privacy of this personal data is an important aspect in this context; however, this is out of scope as we are using a publicly published dataset in this study.

The crucial research questions here are:

1) How far can we support the user in controlling and improving their user model?

2) How will we use the user feedback loop for improving the model?

3) How can we prioritize this data in order to get the best results of the model?

4) What is the level of detail in the model that we should consider?

We can observe that our proposed model is founded on three dimensions: 1) user modeling using machine learning, 2) user control, and 3) ontology-based domains as input for the model.

The overarching research question is: How far can we merge these three dimensions in order to construct a dynamic and controllable user model?

IV. METHODOLOGY AND DESIGN OF THE SYSTEM

This section discusses the methodology approached in this research. Then, it mentions the technologies used in the case study.

A. *Methodology*

Generally, building predictive data analytics solutions for this kind of problems involves a lot more than choosing the right machine learning algorithm. One of the most frequently used methodologies for this is the CRoss Industry Standard Process for Data Mining (CRISP-DM) [27]. The six key steps of the predictive analytics project lifecycle that are defined by the CRISP-DM are; problem understanding, data understanding, data preparation, Modeling, evaluation, and deployment. These steps are expanded in the context of this work below.

Step 1: Problem Understanding

The target is to build a personalized predictive data analytics model that provides services to number of users according to their general interests. These preferences are collected from their email messages. The system can read the email documents, build and train a model, and hence decide whether this person is interested in a certain topic.

The emails of participants are explored and scanned in order to find topics of interest for a person from his email history. Then, email messages are then classified using the appropriate machine learning technique.

Step 2: Data Understanding

It is critical to find the right data in order to be able to solve the problem in hand. In this research, we are working on Enron email dataset [24], which will be described later in detail. It is a huge dataset that contains about 0.5M messages. This step is concerned with selecting the subset of all available data that we will be working with.

There is always a strong desire for including all data that is available, that the maxim "more is better" will hold. We need to consider what data we really need to address the problem in hand. We have to be more disciplined in our data selection then handling to achieve more consistent and accurate results that are likely to attain.

Step 3: Data Preparation

After selecting the data, we needed to study how we are going to use the data. This preprocessing phase is mainly about getting the selected data into a form that we can work. The data preprocessing is divided into three stages; formatting, cleaning, and sampling:

- Formatting: We have selected a format that is suitable to work with. All the mail messages are now in comma separated file format.
- Cleaning: Quality data is a prerequisite for quality predictive models. So, to avoid "garbage in, garbage out" and improve data quality, we have to work this crucial step carefully. Sometimes we find some data instances that are incomplete and do not carry the data there should be. So, we removed all the records of incomplete, noisy, or inconsistent data. In our research, there is a mandatory task which is text preparation. Text cleaning phase includes stripping whitespace, removing stop words, numbers, punctuation, URLs, and links.
- Sampling: In working on this huge data dataset (Enron emails dataset), sampling is an important factor that we have to take into consideration at least in the early learning stages. We have to experience the performance before deciding whether we need to carry on this step. We think that we can start with working on the whole dataset. If we found out that this is inefficient, then we could use a smaller representative sample of the selected data. The target of this step is to be much faster in exploring and prototyping solutions instead of considering the whole dataset.
- Data Transformation: This step is also referred to as feature engineering. As per the problem we are targeting, we are concerned mainly with the text included in the data. we are principally working on analytics for the text data in the users' email messages. Data transformation here is mainly text normalization which means converting it to a more convenient and standard form. For example, most of what we are going to do with language rely on first separating out or tokenizing words from running tokenization text, the task of tokenization. Another part of text normalization is Stemming which refers to a simpler version of lemmatization, in which we mainly strip suffixes from the end of the word.

Step 4: Modeling

The Modeling phase of the CRISP-DM process is when the machine learning work occurs. Different machine learning algorithms are used to build a range of prediction models from which the best model will be selected. The knowledge of the problem domain will influence this step and we will very likely have to be revisited to achieve the optimal solution for the problem in hand.

Step 5: Evaluation

This stage covers all the evaluation tasks required to show that a prediction model will be able to make accurate predictions after being used and that it does not suffer from overfitting or underfitting. This step would be clearer through different experiments in this research journey as this is vague in this stage of study.

Step 6: Deployment

Eventually, the last phase of CRISP-DM covers the work done to successfully integrate a machine learning model into the process within an organization. This phase is not applicable in our research.

The second part of this research is to feed the user model with a wide range of words that enriches it and help it for better understanding the user. This enhancement will improve the prediction accuracy of the model and get better results.

B. *Technologies*

- *Enron Email Dataset*

Enron email dataset was primarily gathered and prepared by the CALO Project (Cognitive Assistant that Learns and Organizes) [28]. It was formerly posted to the web and made public by the Federal Energy Regulatory Commission [24]. It includes a huge amount of data. It is corpus for about 150 persons, which mostly senior managements of Enron. The dataset contains a total of about 0.5M messages that are organized into folders.

There are some modifications done on the original data collected. All attachments were removed. Some messages have been deleted. The deletion was part of a reduction effort that was requested by some employees. There were some invalid email addresses converted to something different. For example, when no recipient was specified, user@enron.com is converted to no_address@enron.com [24].

- *Development*

  In this work, we tended to find integrated machine learning tools and programming languages. R and Python were primarily popped up in our search space due to the availability of multiple open source machine learning libraries. Python is selected as the core programming language including NumPy, SciPy, and Python visualization tools.

  SciKitLearn API is a free software machine learning library. It provides an implementation of a wide range of machine learning algorithms and data preparation methods. SciKitLearn is a simple and efficient tool for data mining and data analysis. It is built on NumPy, SciPy, and matplotlib. This is in addition to that it is open source and commercially usable (BSD license) [29].

V. Progress to Date and Plan of Further Research

To build the target user predictive model and to be able to evaluate it, we applied the model in a usecase, then we divided the journey to number of phases.

Phase 1: One user only

This is the current experiment we are working on. The target of this part of research is to work on the email dataset for one user only and build a model that can predict the points of interest for this user. And at an incoming message, the model can predict whether the user would be interested in this message or the message needs an action from them. And then, it takes the user's feedback and retrains the model to enhance its prediction accuracy. The early results of this experiment show promising prediction accuracy results; however, it still needs more work in the evaluation phase to ensure reliable results.

Phase 2: Classifying the whole dataset

The next phase is to build a model on the whole dataset, the model should be trained on the whole data, and then it can classify the users according to their preferences. At an incoming message for any of the users, the model could predict whether it would be interesting to the user. And then, the users' feedback is gathered and employed to retrain the classification model. This feedback should enhance the model prediction accuracy.

Phase 3: Providing services from Ontology-based domain

The last stage of this study will work on the model built in the previous step. And then we will try to feed the model in a certain domain with words that describe this domain (vocabulary alternatives, different subject, etc.). For example, if we are concerned with football, we have to search for the words football, David Beckham, Manchester United, etc. In such case, we will take these words from an Ontology concerned with football (DBPedia for example). Then, this will feed the model with a big range of words that helps it to understand more about the users and hence improve the results of the model.

VI. Conclusion

Generally, the main goal of personalizing a system is to provide the right information and services to meet user's individual preferences. Recently, it has become increasingly important to deliver not just the right information, but to do so in a highly dynamic way. This makes the environment more reliable and tailored to the change in the user preferences. The other important feature that this work is trying to offer is to incorporate user control in the feedback cycle in machine learning-derived user modeling. This has an advantage over some classical interpretations of personalization.

In this study, we are employing machine learning techniques to achieve this goal. Enabling the user control over a machine learning-derived model is a significant research challenge. This is the main challenge addressed in this early stage of this research. In this paper, we discussed this research idea, current plan, and progress to date.

At this stage of the research there are three core concerns. First, to what extent can we support the user in controlling and improving the model? Future work will require examining user intervention in building a dynamic user model using machine learning.

Secondly, where should the user feedback dimension be placed in the modeling cycle? This will involve identifying a balance between the model intelligence whilst maintaining the scrutability in a machine learning-derived user model.

The third challenge is concerned with the model evaluation and how to evaluate the efficacy, efficiency and reliability of the proposed system.

Acknowledgment

This research is enhanced by ADAPT Centre at Trinity College Dublin. ADAPT Centre for Digital Content Technologies is funded under support from Science Foundation Ireland (SFI) Research Centres Programme and co-funded under the European Regional Development Fund. Special thanks to Prof. Owen Conlan for his continuous guidance, support, and fruitful advice.

References

[1] C. Niederée, A. Stewart, B. Mehta, and M. Hemmje, "A multi-dimensional, unified user model for cross-system personalization," in *Proceedings of the AVI 2004 Workshop on Environments for Personalized Information Access*, 2004, pp. 34–54.

[2] S. McBurney, N. Taylor, H. Williams, and E. Papadopoulou, "Giving the user explicit control over implicit personalisation," in *Procs. of Workshop on Intelligent Pervasive Environments (under AISB'09), Edinburgh, Scotland*, 2009.

[3] D. Ralph and S. Searby, Location and personalisation: delivering online and mobility services, vol. 8. IET, 2004.

[4] O. J. Postma and M. Brokke, "Personalisation in practice: The proven effects of personalisation," *J. Database Mark. Cust. Strateg. Manag.*, vol. 9, no. 2, pp. 137–142, 2002.

[5] G. Fischer, "User modeling in human--computer interaction," *User Model. User-adapt. Interact.*, vol. 11, no. 1, pp. 65–86, 2001.

[6] A. D. A. Daniela Petrelli and G. Convertino, "A User-Centered Approach to User Modeling," in *UM99 User Modeling: Proceedings of the Seventh International Conference*, 2014, vol. 407, p. 255.

[7] M. E. Müller, "Learning scrutable user models," 2002.

[8] M. E. Muller, "Can user models be learned at all? Inherent problems in machine learning for user modelling," *Knowl. Eng. Rev.*, vol. 19, no. 1, pp. 61–88, 2004.

[9] D. Billsus and M. J. Pazzani, "User modeling for adaptive news access," *User Model. User-adapt. Interact.*, vol. 10, no. 2–3, pp. 147–180, 2000.

[10] Y. Yang, M. H. Williams, L. M. MacKinnon, and R. Pooley, "A service-oriented personalization mechanism in pervasive environments," in *Web Intelligence, 2005. Proceedings. The 2005 IEEE/WIC/ACM International Conference on*, 2005, pp. 132–135.

[11] D. Buhalis and A. Amaranggana, "Smart tourism destinations enhancing tourism experience through personalisation of services," in *Information and Communication Technologies in Tourism 2015*, Springer, 2015, pp. 377–389.

[12] O. Conlan, A. Staikopoulos, C. Hampson, S. Lawless, and I. O'keeffe, "The narrative approach to personalisation," *New Rev. Hypermedia Multimed.*, vol. 19, no. 2, pp. 132–157, 2013.

[13] I. O'Keeffe *et al.*, "Personalized activity based eLearning," *Proc. 12th Int. Conf. Knowl. Manag. Knowl. Technol. - i-KNOW '12*, p. 1, 2012.

[14] A. F. Monk and J. O. Blom, "A theory of personalisation of appearance: quantitative evaluation of qualitatively derived data," *Behav. Inf. Technol.*, vol. 26, no. 3, pp. 237–246, 2007.

[15] H. Fan and M. S. Poole, "What is personalization? Perspectives on the design and implementation of personalization in information systems," *J. Organ. Comput. Electron. Commer.*, vol. 16, no. 3–4, pp. 179–202, 2006.

[16] D. Godoy and A. Amandi, "User profiling in personal information agents: a survey," *Knowl. Eng. Rev.*, vol. 20, no. 4, pp. 329–361, 2005.

[17] T. Joerding, "A temporary user modeling approach for adaptive shopping on the Web," in Proceedings of Second Workshop on Adaptive Systems and User Modeling on the World Wide Web, Toronto and Banff, Canada. Computer Science Report, 1999, pp. 7–99.

[18] E. Rich, "Users are individuals: individualizing user models," *Int. J. Man. Mach. Stud.*, vol. 18, no. 3, pp. 199–214, 1983.

[19] A. Kobsa, "Generic user modeling systems," *User Model. User-adapt. Interact.*, vol. 11, no. 1, pp. 49–63, 2001.

[20] J. Hothi and W. Hall, "An evaluation of adapted hypermedia techniques using static user modelling," in *Proceedings of the second workshop on adaptive hypertext and hypermedia*, 1998, pp. 45–50.

[21] M. A. Girolami and A. Kabán, "Simplicial Mixtures of Markov Chains: Distributed Modelling of Dynamic User Profiles.," in *NIPS*, 2003, vol. 16, pp. 9–16.

[22] M. Assad, D. Carmichael, J. Kay, and B. Kummerfeld, "PersonisAD: Distributed, active, scrutable model framework for context-aware services," *Pervasive Comput.*, pp. 55–72, 2007.

[23] A. Staikopoulos *et al.*, "AMASE : A Framework for Composing Adaptive and Personalised Learning Activities on the Web," pp. 190–199, 2012.

[24] W. W. Cohen, "Enron Dataset," 2015. [Online]. Available: https://www.cs.cmu.edu/~./enron/. [Accessed: 12-Jun-2017].

[25] J. Ahn, P. Brusilovsky, D. He, J. Grady, and Q. Li, "Personalized web exploration with task models," in *Proceedings of the 17th international conference on World Wide Web*, 2008, pp. 1–10.

[26] P. De Bra, G. Houbeny, and H. Wu, "AHAM: A Dexterbased Reference Model for Adaptive," in *Hypermedia". Proceedings of the ACM Conference on Hypertext and Hypermedia*, 2009.

[27] J. D. Kelleher, B. Mac Namee, and A. D'Arcy, "Fundamentals of Machine Learning for Predictive Data Analytics." MIT Press, 2015.

[28] "The PAL Framework." [Online]. Available: https://pal.sri.com/. [Accessed: 12-Jun-2017].

[29] F. Pedregosa and G. Varoquaux, "Scikit-learn: Machine learning in Python," ... *Mach. Learn.* ..., vol. 12, pp. 2825–2830, 2011.

Complexity Reduction in Graphs: A User Centric Approach to Graph Exploration

Tim Grube, Florian Volk, Max Mühlhäuser
Telecooperation Group, TU Darmstadt, Germany
Email: {grube, volk, max} @tk.tu-darmstadt.de

Suhas Bhairav, Vinay Sachidananda, Yuval Elovici
iTrust, SUTD, Singapore
Email: {suhas_setikere, sachidananda, yuval_elovici} @sutd.edu.sg

***Abstract*—Human exploration of large graph structures becomes increasingly difficult with growing graph sizes. A visual representation of such large graphs, for example, social networks and citational networks, has to find a trade-off between showing details in a magnified view and the verall graph structure. Displaying these both aspects at the same time results in an overloaded visualization that is inaccessible for human users. In this paper, we present a new approach to address this issue by combining and extending graph-theoretic properties with community detection algorithms. Our approach is semi-automated and non-destructive. The aim is to retain core properties of the graph while–at the same time–hiding less important side information from the human user. We analyze the results yielded by applying our approach to large real-world network data sets, revealing a massive reduction of displayed nodes and links.**

Keywords*—Complexity reduction; graph visualisation; big data exploration; graph metrics; community detection.

I. Introduction

Computing on data sets is becoming increasingly easier. With the rise of *big data* and powerful computing devices, the collection and processing of large data sets has become a common thing. Research as well as industry profits a lot from this capability to reveal new insights and connections that can only be detected by analyzing large amounts of data.

However, Moore's Law [1] does not apply to the ability of human users to understand and explore such big data sets. Making large data sets, for example, networks, accessible to human users is difficult and becomes increasingly more difficult with ever-increasing data sets. Human-centric data analysis techniques usually employ visualization of the data sets that the human user intends to work with. Visualizing real-world networks as connected nodes quickly results in an inaccessible chaos due to large amount of information to be shown. See Figure 1 for a visualization of a part of the Facebook social network [2]. In order to comprehend relations within such a network, a user needs to magnify the visualization a lot. This magnification implies a loss of overview, so that a user might be able to understand specific relations but loses track of the overall structure of the data set.

This issue of simultaneous visualization of details and overall structure complicates the process of exploring data sets when the outcome of the exploration is not pre-defined. Human users are often involved in exploring data sets when the aim is to find new insights, connections and cross-references. Pre-defined and well-specified analyses can be processed automatically without human involvement, and thus, no need for a

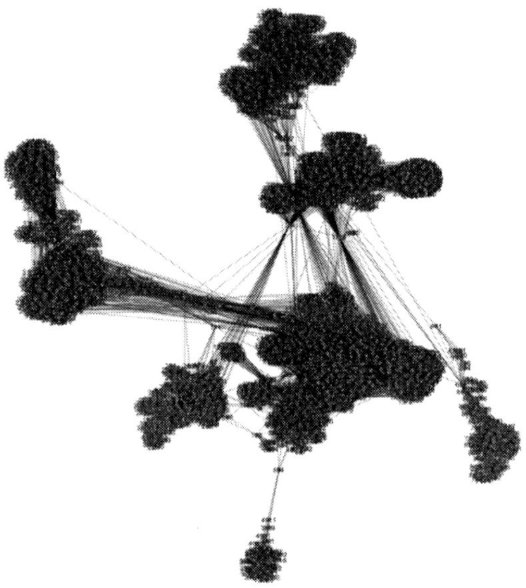

Fig. 1. User relations in a social network (data taken from Facebook).

visualization. However, exploratory data analysis usually does not allow for automated pre-filtering of data sets to obtain a human-centric, *decluttered* view on the data.

The technique we present in this article combines and extends graph-theoretic properties with community detection. This technique aims at reducing the visual complexity of network data sets in order to render these more accessible to human users. For this reduction of visual complexity, we propose a semi-automated, non-destructive approach to identify core insights, while side-information is hidden.

Our results indicate a massive reduction of displayed nodes and links in every iteration of our proposed approach. Therefore, only one to three iterations effectively reduce graphs as the one shown in Figure 1 to a representation that is easy to comprehend for human users (see Figure 3 for the results of only three iterations).

The remaining paper is structured as follows: in Section II, we discuss relevant related work on complexity reduction in graphs; in Section III, we introduce fundamental background information to our contribution by discussing used graph theory measures and community detection mechanisms. We continue in Section IV with the introduction of our technique of reduce complexity and we present our findings in Section V. We conclude this paper in Section VI with a summary and future research directions.

II. Related Work

In this section, we discuss related work with respect to the field of complexity reduction in graphs.

Kimelman et al. [3] proposed techniques like ghosting, hiding and grouping of edges. The nodes and edges of the graph were removed based on various techniques like weights of edges, labels of nodes, etc., but these techniques were concerned with dynamic graphs. Differently, Holten et al. [4] reduced only the visual cluttering of edges by bundling them.

Fisheye techniques [5], [6] tend to concentrate only on the interesting regions of the user. The zooming feature in such techniques is only responsible for making a very small region of a graph appear larger. They do not remove any nodes or edges. So, the overall graph content remains the same. Fisheye views that retain structure are introduced by Furnas [7]. Abello et al. [8] introduced hierarchical clustering and depiction of a treemap in addition to a compound fisheye view technique but never concentrated on reducing the overall size of a network.

Various approaches towards creating communities in large graphs are presented in [9]–[11]. These techniques provide a significant level of understanding of the kind of nodes and their properties in large networks but never used the same to reduce the overall content in a large network and provide a simpler view.

Sundararajan et al. [12] introduced Rectangular Partitioning and Voronoi Partitioning techniques. The former involved partitioning the area of display into four quadrants while the latter involved the partition area being closer to the concerned node. This only reduces the distortion in the graphs.

Batagelj et al. [13] took a mathematical approach through the usage of matrix. Large graphs were reduced to *k-cores*. Later, the graphs are represented as an adjacency matrix or a contextual matrix based on their size. But when the graphs grows really large, managing the matrix becomes a humongous task.

All before mentioned techniques reduce complexity based on a global perspective, disregarding the current interests of a user. Thus, these techniques may maintain and highlight information that is not relevant the user's current situation while rejecting important information from the user's slant.

III. Background

In this section, we introduce our terminology that is used to model data sets using graphs. After this terminology description, we introduce the foundation of various graph-theoretic properties and discuss them in the context of our complexity reduction. Additionally, we introduce those community detection algorithms that we use to reduce the visual complexity.

A. Terminology: Graph-based Data Representation

Many data sets can be represented by a Graph $G = (V, E)$ that is formed by a set of *vertices* $v_i \in V$, which represents the pieces of data. Relations and *connections* between these information are represented by *edges* $E \subseteq V \times V$, i.e., connections are represented by pairs of vertices (v_x, v_y). Considering our exemplary application scenario of a social network, each vertex represents a user and edges represent the connections between users. Another example is the model of (research) citations as citation graph where publications are represented as vertices, and a citation is represented by a directed edge.

A graph G can be *directed* or *undirected*. In a directed graph, an edge e_k can exist between a vertex v_i and v_j while the other direction (e_l: v_j to v_i) may or may not exist, independent of the existence of e_k. In an undirected graph, the existence of before-mentioned edge e_k also implies the existence of e_l. A social network like Facebook applies undirected connections, thus, if user Alice is connected to user Bob, Bob is also connected to Alice (Alice and Bob are "friends"); in opposite to that, Twitter applies directed connections. Thus, Alice may be connected to Bob (Alice "follows" Bob), but Bob may not be connected to Alice.

The number of connections of a vertex v_i is noted as the *degree* d_{v_i} of v_i. In case of a directed graph, the degree of a vertex has to be specified for outgoing connections: the *out-degree* and for incoming connections: the *in-degree*.

A connection between any two vertices is called a path p. A path between adjacent vertices has the length 1; yet, a path may include intermediate vertices to connect them. A path p is represented by an sequence of edges $p = (e_0, e_1, ..., e_n)$. The *shortest path* between two vertices is the path with least edges.

B. Selection Criteria Values

Vertex selection is one of the fundamental steps in reducing the complexity of a graph, i.e., the selection of an appropriate subgraph. Selecting a set of vertices based on specific parameters helps in creating a subgraph that retains its inherent properties and reflects the user's interests. The vertices are identified based on *selection criteria values* (SCV). A SCV acts as the basis of vertex selection in our framework. For a vertex to be selected as part of a graph, its SCV must be greater than the SCV of the user's interest. Here, we express the user's interest by selecting a start vertex whose SCV is compared with every other vertex in the network while reducing the graph. The SCVs include graph-theoretic properties and centrality measures but are not limited to just these presented measures. We focus this paper on the usage of importance, connectivity, and distance measures as these appeared most promising in our literature research. The importance measures are PageRank and Betweenness Centrality; the connectivity measures are Clustering Coefficient and Degree Centrality; the distance measures is Closeness Centrality. In the following sections, we discuss every SCV to understand their significance to our idea of reducing visual complexity.

1) Closeness Centrality: Closeness Centrality determines the closeness of vertices in a network, i.e., it determines the distance of vertices in a graph. The vertices that have high closeness centrality values are considered to be closer to other vertices in the graph. For example, in a network where information flows from one vertex to another, transmission of information takes place quickly due to their high closeness centrality values.

According to Freeman [14], closeness centrality of a vertex is defined as:

"The sum of graph-theoretic distances from all other vertices, where the distance from a vertex to another is defined as the length of the shortest path from one to the other."

Closeness is generally attributed to the quickness in flow of information in the network. The quicker the information arrives at or departs from a vertex, the closer is the destination or the source vertex respectively [15].

In a social network scenario, this could mean when an information is shared between two individuals, the chances of such information propagating in the network is higher with those who are immediate friends or neighbors. For a given vertex p_i, closeness centrality [16] is calculated using (1), where V denotes the total number of vertices in the graph, i and k are integers where $i \neq k$, $d(p_i, p_K)$ denotes the number of edges in the shortest path between p_i and p_k.

$$C_C(p_i) = \frac{V-1}{\sum_{k=1}^{V} d(p_i, p_k)} \tag{1}$$

2) Betweenness Centrality: Betweenness Centrality depicts influence or powerfulness of a vertex in a network. The vertices that have high betweenness centrality values are *highly influential* as they act as bridges for interactions between several pairs of vertices. In a social network scenario, where individuals form friendships with individuals who in turn have a large group of friends tend to have an influential clout due to their connections with such individuals.

According to Freeman [14], the betweenness centrality of a vertex is defined as:

"The share of times that a vertex i needs a vertex k (whose centrality is being measured) in order to reach a vertex j via the shortest path."

Betweenness centrality [16] can be calculated using (2), where g_{jk} denotes the total number of shortest paths between p_i and p_k and $g_{jk}(p_i)$ denotes the number of shortest paths containing p_i.

$$C_B(p_i) = \sum_{i \neq j \neq k \in V} \frac{g_{jk}(p_i)}{g_{jk}} \tag{2}$$

3) PageRank: PageRank [17] makes use of the idea that was conceived to compute citations to a web page and also add its own Midas Touch by ensuring that all web pages are not treated equally but by applying normalization to the web page depending on the total number of links present on that page. In this way, PageRank provides a rank to every web page.

A transition matrix is created based on the transfer of importance from one to another. Initially, we apply uniform distribution based on the initial grade structure. Then, depending upon the incoming links we re-calculate the PageRank value [18].

The vertices that have high PageRank values are considered more important due to their high incoming links. For example, in the world wide web network, a web page that is referenced in many other web pages will have a high PageRank value. In a social network scenario, individuals with higher PageRank are generally highly powerful due to a lot of connections whose betweenness centrality value is also high. Since the concept of PageRank is based on Betweenness Centrality, the PageRank value of a vertex depends on its betweenness centrality value and its neighbors.

4) Degree Centrality: Degree Centrality portrays the level of connectivity of a vertex in a network. The degree centrality is computed as presented in (3) and given by the number of adjacent neighbors. The vertices that have high degree centrality values are more influential or important. For example, in a social network, a vertex with a high number of connections is powerful and highly visible when compared to others [10].

$$C_D(p_i) = deg(p_i) \tag{3}$$

5) Clustering Coefficient: Clustering Coefficient gives an indication of clusters in a network. The vertices that have high clustering coefficient values are more closely knit together. For example, in a social network, vertices with high clustering coefficients have a desire to form close bonds or friendships with their neighbors [19].

According to Zafarani et al. [20], the clustering coefficient can be defined as shown in (4).

$$CC = \frac{3 \times (Number\ of\ triangles)}{Number\ of\ Connected\ Triples\ of\ Vertices} \tag{4}$$

C. Community Detection Algorithms

Community detection allows to group vertices that share similar properties. The properties vary depending on the type of the used community detection algorithm (CDA). For example, Louvain algorithm optimizes communities with respect to maximizing their modularity while direct K-means community detection optimizes on the spread of communities.

We use the found communities to reduce complexity in larger steps, being able to remove larger portions of the graph in the first few complexity reduction cycles. CDAs form one of the core evaluation criteria in our implementation. The four algorithms used in our framework are Hierarchical Clustering, Original Louvain, Louvain with Multilevel Refinement, and Direct Clustering. In the following sections, we will discuss the CDAs that help us in understanding the way in which various clusters are formed.

1) Hierarchical Clustering: The hierarchical clustering method by Jain and Dubes [21] involves obtaining a series of partitions that are nested by transforming a proximity matrix. We utilize the complete-link clustering method in an agglomerative setting in our implementation. One of the main advantages of this method is the avoidance of chaining-effects. Also, the clusters are balanced and smaller when applying this method.

2) Original Louvain: The Original Louvain algorithm by Blondel et al. [22] aims to increase the graph property *modularity* [23] due to reassigning vertices from one community to another one [24]. A local moving heuristic is used to optimize the modularity graph property in iterative steps, i.e., each vertex evaluates if moving to a neighbor's community can increase the modularity and reassigns itself to the modularity maximizing community. After that, a reduced size graph is established, where the vertices are the previously established communities.

3) Louvain with Multilevel Refinement: Louvain with Multilevel Refinement algorithm was formulated and conceived by Rotta et al. [25]. The algorithm is similar to Original Louvain except for the fact that the local moving heuristic is applied twice. Louvain with Multilevel Refinement generates community assignments yielding higher modularity values compared to the Original Louvain.

4) Direct K-means: The direct K-means algorithm by Alsabti et al. [26] partitions a data set into k communities where k is fixed. First, k vertices are selected randomly, and the remaining vertices are assigned to the community of the closest of the k selected vertices. After assigning all vertices, the means of the communities are computed and assigned to the new k selected start nodes for the next iteration. This process is repeated until stabilization.

IV. Complexity Reduction Approach

In this section, we present our framework to reduce the data complexity in graphs.

A. Subjectivity of Visual Complexity

Visual complexity is a subjective measure. Every user has a different understanding of visual complexity of a data set, depending on previous experience, expert knowledge, and familiarity with the respective subject. Hence, the challenge of visual complexity reduction cannot be solved with an "one for all" approach.

To overcome this subjectivity challenge, we provide a round-based complexity reduction method that allows to reduce the visual complexity in succeeding iterations until the user is able to understand the data set well enough to collect the desired information.

```
1: function REDUCEVISUALCOMPLEXITY(g, r, s, id, type_g, c_alg)
     g: Input graph
     r: Number of iterations
     s: Selection criteria
     id: Start node (picked by user or at random)
     type_g: Graph Type (directed or undirected)
     c_alg: Community detection algorithm
     g_reduced: Output graph with reduced complexity

2:     g_reduced ← new Graph();
3:     for i ← 1 to r do
                                          ▷ loc: List of Communities
4:         loc ← getCommunities(id, g, algo, g_type)
5:         for j ← 1 to sizeOf(loc) do
6:             community ← loc.get(j)
7:             for k ← 1 to sizeOf(community) do
8:                 v_k ← community[k]
9:                 if getSCV(v_k, s) > getSCV(id, s) then
                                          ▷ compute shortest path
10:                    sp ← getSP(v_k, id)
11:                    addToReducedGraph(sp, g_reduced)
                          ▷ break if applied w/ community detection
12:                    if thenloc > 0
13:                        break loop
14:                    end if
15:                end if
16:            end for
17:        end for
18:        i++
19:    end for
20:    return g_reduced
21: end function
```

Fig. 2. Iterative Reduction of Visual Complexity

B. Visual Complexity Reduction

Our method to reduce the visual complexity is a two-step process, which also allows to parallelize the computation to reduce computation time faced by the user. We give an algorithmic description in Algorithm 2. In the first step, we apply a community detection algorithm to group similar information. According to lines 3-6 of Algorithm 2, we compute the disjoint communities and proceed with each of them individually. At this point, we can easily parallelize the computation, having the communities being handled concurrently.

In the second step, we have to differentiate whether community detection was applied in the first place or not. If community detection is not applied, we calculate the SCV of each vertex and compare it with the SCV of the vertex representing the user's interest. When the SCV of the inspected vertex is higher than a threshold, which is in our case the SCV of the user's interest, the vertex is retained in the reduced graph. To preserve the relation and connection of the user's interest and inspected vertex, we retain not only the vertex but

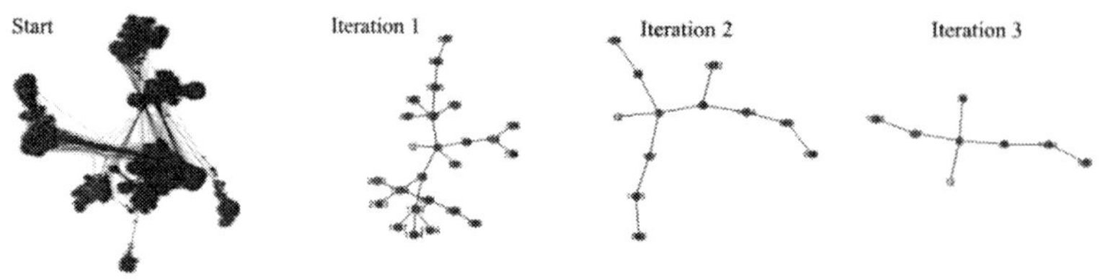

Fig. 3. Reduction of graphs by applying Algorithm 2 (first four iterations shown).

also the vertices on the shortest path. In our social network scenario, these vertices represent the chain of friends between two subjects and, thus, may be relevant to the user.

If community detection was applied in the first place, we only retain a single representative of each community and hide the remaining vertices of a community. This way, we can easily achieve a massive complexity reduction in only a few iterations. Yet, the user can select the representative of a community as next interest, restarting the process from this new perspective and gain new insights into the data set.

These two steps are then repeated until the user is able to sufficiently comprehend the graph. Our algorithmic description uses a predefined number of iterations, yet an application would delegate this decision to the user, who might request further iterations.

In Figure 3, we visualize the iterative complexity reduction process. On the left-hand side, the original graph, which fed into the visual complexity reduction mechanism, is visualized. In three steps, visualized towards the right-hand side, we reduce the visual complexity of the data set considering a randomly chosen vertex to reflect an otherwise user-selected interest. The user interest is marked by the green vertex with the ID 12 in Figure 3. The original data set, labeled "Start", is the Facebook data set already presented in the introduction in Figure 1.

V. EVALUATION

In a simulation study, we evaluated the performance and effectiveness of our proposed technique to reduce the visual complexity of data sets. In the next section, we provide insights into our experimental setup and details of our simulation study. After that, we present and discuss our results for different network classes.

A. Experimental Setup

In this section, we describe and discuss our selection of data sets, followed by the description of our simulation setup.

a) Networks: To provide useful insights, we decided to perform our simulation based on samples of real-world networks. We imagine the usage of our technique in applications based on social networks (with respect to to social ties), in (scientific) libraries (with respect to to co-authorship), biological data (with respect to protein interactions), and computer networks (with respect to to the logical interconnections) as described in Table I. We restrict our simulation to the largest connected component of the data set, as accounting for importance and relationship of non-related information is out of scope of this paper.

TABLE I
LARGE NETWORK DATA SETS

Class	data set	Vertices	Connections
Social	Facebook (FB)	4,029	88,234
Library	Arxiv gr-qc (CA)	4,158	13,428
Biological	Vidal (VI)	3,133	6,726
	Moreno (MO)	1,870	2,277
Computer	p2p-Gnutella08 (P8)	6,301	20,777

As the representative for a social network, we decided to use a sample of Facebook's social graph [2]. This data set (FB) consists of 4,029 vertices and 88,234 connections, representing users of Facebook and their respective interconnections.

As the representatives for library-based and collaboration-based data sets, we selected a co-authorship data set [2] that reflects the research collaborations and co-authorships in the Arxiv gr-qc (General Relativity and Quantum Cosmology) area. This data set (CA) consists of 4,158 vertices and 13,428 connections in their largest connected component.

As the representatives of biological networks, we selected the protein interaction networks vidal [27] and moreno [28]. These data sets reflect interactions of proteins on a molecular level. The vidal data set (VI) consists of 3,133 vertices and 6,726 connections; the moreno data set (MO) consists of 1,870 vertices and 2,277 connections.

As the representative of computer networks, we selected a sample of the Gnutella network [2] that reflects the interconnections of users in the Gnutella network. The sample (P8) is sampled on August 8, 2002 (p2p-Gnutella08) and consists of 6,301 vertices and 20,777 connections.

b) Simulation Setup: We perform a reduction of visual complexity on each of the aforementioned data sets and calculate graph-theoretic properties that are also used as SCVs, namely closeness centrality (C_C), betweenness centrality (C_B), PageRank (PR), degree centrality (C_D), and clustering coefficient (CC).

We perform the complexity reduction on each of the data sets with each pairwise combination of selection critera values (SCV) and community detection algorithms (CDAs) been described in Sections III-B and III-C and compare and interpret

TABLE II
SIMULATION DETAILS

Property	Value(s)
SCV	Closeness Centrality (C_C) Betweenness Centrality (C_B) PageRank (PR) Degree Centrality (C_D) Clustering Coefficient (CC)
CDA	Hierarchical Clustering (HC) Direct Clustering (DC) Original Louvain (OL) Louvain w/ multilevel refinement (LM)
Repetitions	15
Iterations	3
User Interest	V.getRandomVertex()

the results represented by the statistical mean values and number of removed vertices.

We need a user interest expressed as one "preselected" piece of information from the data set to apply our technique to reduce visual complexity. To account for this, we selected a random vertex as user interest and repeat this whole process 15-times.

In Table II, we summarize the details of our simulation study and used abbreviations.

B. Results

In this section, we present the results of our simulative study. The results are structured according to the class of networks as aforementioned. Detailed measurements are shown in Table III and IV.

a) Social Networks: Social networks, here the FB data set, exhibit a strongly connected core, the so-called *rich club*. These nodes are connectors between various clusters. This structural behavior leads to comparable short paths, i.e., the closeness centrality is comparably high, and core vertices exhibit a high betweenness and degree centrality. Moreover, due to the "clustering" of vertices, the clustering coefficient is also high (in the FB data set: 0.6055).

Our complexity reduction technique reduces up to 4,035 (99.9%) vertices in the three iterations. All combinations but one of the SCV and CDA combinations preserve the core of the data set which is indicated by the high C_B, C_C, and C_D values. These indicate that the retained information is representative for their clusters. The drop of PR and CC indicate that the preserved structure roughly follows a star-topology, which is also shown in Figure 3.

Using CC as only SCV leads to a different result where different information is retained, yet, this is expected. The core vertices have by definition a lower CC value and are, thus, not likely to be above the threshold. These vertices are connected to a multitude of different other vertices that are forming their own, smaller clusters.

b) Library/Collaboration Data Set: A collaboration, or library, data set yields a similar structure to a social network. Collaborators establish clusters and an inner core of highly influential people and publications. The gr-qc data set reveals a high CC (0.5296) and a strongly connected core. While the diameter of the data set suggests otherwise, the core of the data set still shows a high connectivity and tendency towards short paths. That is evident when comparing the diameter (17) and the diameter when only 90% of vertices have to be connected (0.9-percentile effective diameter: 7.6).

Our complexity reduction technique removes up to 4,196 vertices in the three performed iterations. We can see that most combinations of SCV and CDA perform similarly, producing star-topology-like results retaining relevant information of the core of the data set as indicated by higher C_B and C_D. The drop of the CC supports the star-topology again.

However, using CC or C_D as SCV produces different results. Using one of these properties as SCV retains information revealing the significantly higher CC, C_D, and PR and lower C_D values. Thus, retaining representatives of smaller but better and tighter connected clusters.

c) Biological Data Set: Protein interaction networks reveal a different structure, the spread between average and maximum degree is between factor 23 and 30; this factor is by a magnitude smaller than in the previously discussed social networks. This drop in degree goes hand in hand with increased diameter (13–19) and a massive decline in the CC (0.035–0.055 compared to 0.6055 on the Facebook data set).

Our complexity reduction technique removes up to 3,125 (99.74%) vertices on the vidal protein interaction network and up to 1,860 (99.47%) vertices on the moreno protein interaction network. Our technique performs similarly on both data sets, and produces star topologies regardless of the used combination of SCV and CDA. Thus, the resulting C_B is comparably high, and the C_C is very low < 0.0001. The CC also drops to 0.01–0.0004 and endorses the star topology as the remaining vertices are hardly forming local clusters. The PR is similarly low. Yet, the CO-DC combination retains local clusters, which is indicated by higher C_C and CC values.

d) Computer Networks: Computer networks, in this case a sample from the Gnutella network, are resembling either AS-networks with social-like structures on a large scale or, if consisting of only "a few" computers (compared to the whole population), random networks. As the Gnutella network was comprised of only a small subset of all Internet users, the Gnutella sample is akin to a random network. Yet, the degree distribution reveals on the base data set exponentially distributed degrees. Nonetheless, the PageRanks reveal the lack of a highly connected inner core; this core is essential to a "social" network.

Our complexity reduction techniques removed up to 6,299 (99.97%) vertices. The consistently high C_B shows that some well-connected vertices are retained, but the retained vertices are distant from each other, thus, resulting in very low C_C values; only PR used as SCV preserves closer vertices. Using C_C, CC, or PR as SCV results in retaining local clusters in the results while the other SCVs do not keep clusters in their reduced data sets – this is visible by the high CC and PR values for these SCVs.

C. Destructive vs. Non-Destructive Complexity Reduction

The complexity reduction can be performed in two variations: a destructive and a non-destructive one.

First, the execution can be *destructive* by removing vertices that are not matching the user's interest, i.e., removing vertices when their SCV is lower than the threshold and if they are not selected to be a representative. During execution, the data set will shrink and, thus, free memory and reduce the computational complexity of the calculation of SCVs with each performed complexity reduction iteration. However, the user cannot revert the reduction process to earlier stages of complexity reduction.

Second, the execution can be *non-destructive* by, e.g., concealing "removed" vertices only without deleting them from the actual graph. This variation allows moving back and forth between different stages of complexity reduction. Moreover, this also allows a user to adapt her interest, i.e., she can use gained knowledge and re-run the visual complexity reduction with a different focus/interest.

The main functionality of our proposed approach to visual complexity reduction remains unaltered by this design decision as it only affects the ability to move through the different stages of complexity reduction and the visualization for the human user.

Our proposed technique to reduce the visual complexity is capable of removing the vast majority of information in just a few iterations as we have shown before. In order to overcome unwated loss of information during the process of data reduction, our approach always allows the user to track back and select different nodes as user interest. This, however, is infeasible for dynamic data sets, which can be adressed, for example, by caching results from earlier iterations.

VI. Conclusion

The collection and processing of large data sets got common with the rise of *big data* and powerful computing devices. Human users are hardly able to keep up with the increasing pace of collecting and accruing data. Accessing and–more important–understanding these data sets becomes difficult.

In this article, we introduced a combination of graph-theoretic properties and community detection to reduce the visual complexity to render the visualizations of data sets more usable and useful for human users.

Our simulation study has shown that our combination of graph-theoretic properties, measuring the importance of data in the data set, and community detection, grouping similar data in the data set, is able to reduce the visual cluttering of information efficiently. If performed in a non-destructive setting, i.e., if discarded data is only concealed and not removed from the data set, human users can shift their focus when inspecting a data set to account for new insights gained

TABLE III
Properties of Complexity-reduced Data Sets of Facebook, Collaboration, and Gnutella. Protein Interaction Networks.

	Technique	Mean Values				
		C_B	C_C	CC	C_D	PR
FB	C_C & DC	2211000	0.3457	0.00038	298.1	0.00215
	C_C & HC	2356000	0.3478	0.00033	311.9	0.00230
	C_C & OL	1925000	0.3425	0.00037	260.3	0.00191
	C_C & LM	1925000	0.3425	0.00037	260.3	0.00191
	C_B & DC	1453000	0.3327	0.00148	205.9	0.00163
	C_B & HC	1685000	0.3182	0.00604	219.5	0.00209
	C_B & OL	2112000	0.3397	0.00291	251	0.00197
	C_B & LM	2253000	0.3389	0.00356	262.5	0.0021
	PR & DC	1520000	0.3308	0.00268	203.9	0.0017
	PR & HC	1754000	0.3356	0.00208	223.5	0.00178
	PR & OL	1588000	0.3332	0.00284	206.2	0.00163
	PR & LM	1809000	0.3339	0.00336	232.6	0.00185
	C_D & DC	1539000	0.3304	0.0024	219.7	0.00175
	C_D & HC	1810000	0.3291	0.00492	236.4	0.002
	C_D & OL	1943000	0.3311	0.0063	236.2	0.00194
	C_D & LM	2100000	0.3356	0.00451	262.3	0.00211
	CC & DC	70010	0.2638	0.07401	14.13	0.00013
	CC & HC	79330	0.2639	0.08139	15.11	0.00014
	CC & OL	30171	0.27314	2.42765	36.62	2.0282
	CC & LM	83490	0.2641	0.08111	15.71	0.00015
CA	C_C & DC	73470	0.00005	0.02978	16.36	0.00041
	C_C & HC	71590	0.00005	0.04217	8.94	0.00031
	C_C & OL	74020	0.00005	0.02445	17.66	0.00044
	C_C & LM	73470	0.00005	0.02978	16.36	0.00041
	C_B & DC	233900	0.00005	0.00592	16.95	0.00058
	C_B & HC	224500	0.00005	0.00579	16.47	0.00057
	C_B & OL	158000	0.00005	0.01006	14.12	0.00049
	C_B & LM	151600	0.00005	0.00884	13.93	0.00047
	PR & DC	222000	0.00005	0.0015	20.85	0.00067
	PR & HC	98330	0.00005	0.00811	14.92	0.00045
	PR & OL	215300	0.00005	0.00445	19.91	0.00063
	PR & LM	226500	0.00005	0.00266	20.64	0.00066
	C_D & DC	28418	4.44	1.57434	8.38	0.88412
	C_D & HC	28261	4.44	1.57331	8.37	0.88485
	C_D & OL	28370	4.44	1.57288	8.37	0.88744
	C_D & LM	97370	0.00004	0.00991	12.61	0.00039
	CC & DC	22449	4.17	0.84129	4.83	2.43779
	CC & HC	22449	4.17	0.84129	4.83	2.43779
	CC & OL	21060	4.14	0.80191	4.63	2.58936
	CC & LM	43790	0.00003	0.1903	7	0.0003
P8	C_C & DC	38492	0.00023	1.42208	9.55	3.59775
	C_C & HC	39680	0.00023	0.00003	9.53	0
	C_C & OL	38492	0.00023	1.42208	9.55	3.59775
	C_C & LM	38666	0.00023	1.43795	9.54	3.59132
	C_B & DC	82970	0.00023	0	9.92	0
	C_B & HC	88810	0.00023	0.00003	9.72	0
	C_B & OL	162900	0.00023	0	9.84	0
	C_B & LM	137000	0.00023	0	9.83	0
	PR & DC	20920	0.61077	0.58976	3.77	4.02269
	PR & HC	20920	0.61077	0.58976	3.77	4.02269
	PR & OL	18941	0.61688	0.57661	3.69	4.09170
	PR & LM	21934	0.60621	0.59507	3.81	4.14545
	C_D & DC	119200	0.00023	0	12.68	0
	C_D & HC	122900	0.00023	0	12.55	0
	C_D & OL	47420	0.00023	0.00002	10.18	0
	C_D & LM	47420	0.00023	0.00002	10.18	0
	CC & DC	46280	0.00393	2.44391	9.69	3.76677
	CC & HC	46304	0.00393	2.44341	9.69	3.76735
	CC & OL	46280	0.00393	2.44391	9.69	3.76677
	CC & LM	46576	0.00388	2.43269	9.69	3.75996

TABLE IV
PROPERTIES OF COMPLEXITY-REDUCED DATA SETS OF PROTEIN INTERACTION NETWORKS.

	Technique	Mean Values				
		C_B	C_C	CC	C_D	PR
VI	C_C & DC	57420	0.00009	0.00196	15.28	0.00107
	C_C & HC	57420	0.00009	0.00196	15.28	0.00107
	C_C & OL	58990	0.00009	0.00154	15.73	0.0011
	C_C & LM	54320	0.00009	0.00232	14.61	0.00102
	C_B & DC	61750	0.00009	0.00045	15.6	0.0011
	C_B & HC	58410	0.00009	0.00044	15.06	0.00106
	C_B & OL	57790	0.00009	0.00046	14.98	0.00106
	C_B & LM	56350	0.00009	0.00046	14.78	0.00104
	PR & DC	62850	0.00009	0.00600	15.86	0.00112
	PR & HC	138900	0.00009	0.00133	29.01	0.00203
	PR & OL	49980	0.00009	0.00686	13.65	0.00097
	PR & LM	85260	0.00006	0.01049	18.23	0.00136
	C_D & DC	31410	0.00009	0.02806	9.053	0.00064
	C_D & HC	156200	0.00009	0.00314	31.63	0.00221
	C_D & OL	83280	0.00007	0.04043	18.82	0.00136
	C_D & LM	111300	0.00007	0.04704	24.04	0.00172
	CC & DC	23410	8.61693	1.67212	7.98	0.00318
	CC & HC	30480	0.00008	0.1194	8.65	0.00063
	CC & OL	23590	0.00009	0.07014	7.94	0.00057
	CC & LM	24450	0.00009	0.08001	8.03	0.00058
MO	C_C & DC	33615	0.00005	0.00681	6.31	0.00115
	C_C & HC	31886	0.00005	0.00596	6.16	0.00114
	C_C & OL	33467	0.00005	0.00668	6.33	0.00116
	C_C & LM	42355	0.00005	0.0124	7.32	0.00132
	C_B & DC	22275	0.00004	0.00152	4.99	0.00096
	C_B & HC	21245	0.00004	0.00153	4.91	0.00094
	C_B & OL	62989	0.00004	0.00306	9.87	0.00178
	C_B & LM	61556	0.00004	0.00312	9.7	0.00175
	PR & DC	33375	0.00004	0.00595	6.67	0.00125
	PR & HC	30472	0.00004	0.00604	6.34	0.0012
	PR & OL	32441	0.00004	0.0062	6.51	0.00123
	PR & LM	24884	0.00004	0.00618	5.70	0.00109
	C_D & DC	25683	0.00004	0.00277	5.82	0.00109
	C_D & HC	23793	0.00004	0.0028	5.64	0.00106
	C_D & OL	24978	0.00004	0.00284	5.80	0.00108
	C_D & LM	61071	0.00004	0.00376	10.53	0.00189
	CC & DC	20061	0.00004	0.0867	5.28	0.00099
	CC & HC	20061	0.00004	0.0867	5.28	0.00099
	CC & OL	20503	0.00004	0.0857	5.33	0.001
	CC & LM	21185	0.00004	0.0845	5.39	0.00101

by the visual inspection of data sets. The complexity reduction can then be performed again with a focus on new, shifted interests.

Our proposed technique opens an additional direction of research to support user-centric systems in rising exposure to big data and accruing amounts of data. Future research may include user studies to select the–per application field–appropriate graph-theoretic properties in order to achieve an appropriate visual complexity reduction.

ACKNOWLEDGEMENTS

This work has been co-funded by the DFG as part of project B.2 within the RTG 2050 Privacy and Trust for Mobile Users and by the German Federal Ministry for Education and Research (BMBF, Software Campus project KomBi, grant no. 01IS12054).

REFERENCES

[1] G. E. Moore, "Readings in computer architecture," M. D. Hill, N. P. Jouppi, and G. S. Sohi, Eds. San Francisco, CA, USA: Morgan Kaufmann Publishers Inc., 2000, ch. Cramming More Components Onto Integrated Circuits, pp. 56–59.

[2] J. Leskovec and A. Krevl, "SNAP Datasets: Stanford large network dataset collection," http://snap.stanford.edu/data, Jun. 2014.

[3] D. Kimelman, B. Leban, T. Roth, and D. Zernik, "Reduction of visual complexity in dynamic graphs," in *International Symposium on Graph Drawing*. Springer, 1994, pp. 218–225.

[4] D. Holten and J. J. Van Wijk, "Force-directed edge bundling for graph visualization," in *Computer graphics forum*, vol. 28, no. 3. Wiley Online Library, 2009, pp. 983–990.

[5] M. Sarkar and M. H. Brown, "Graphical fisheye views of graphs," in *Proceedings of the SIGCHI conference on Human factors in computing systems*. ACM, 1992, pp. 83–91.

[6] E. R. Gansner, Y. Koren, and S. C. North, "Topological fisheye views for visualizing large graphs," *IEEE Transactions on Visualization and Computer Graphics*, vol. 11, no. 4, pp. 457–468, 2005.

[7] G. W. Furnas, "Generalized fisheye views," in *Proceedings of the SIGCHI Conference on Human Factors in Computing Systems*, ser. CHI '86. New York, NY, USA: ACM, 1986, pp. 16–23.

[8] J. Abello, S. G. Kobourov, and R. Yusufov, "Visualizing large graphs with compound-fisheye views and treemaps," in *International Symposium on Graph Drawing*. Springer, 2004, pp. 431–441.

[9] A. Clauset, M. E. J. Newman, and C. Moore, "Finding community structure in very large networks," *Phys. Rev. E*, vol. 70, p. 066111, Dec 2004.

[10] S. P. Borgatti, A. Mehra, D. J. Brass, and G. Labianca, "Network analysis in the social sciences," *science*, vol. 323, no. 5916, pp. 892–895, 2009.

[11] M. Girvan and M. E. Newman, "Community structure in social and biological networks," *Proceedings of the national academy of sciences*, vol. 99, no. 12, pp. 7821–7826, 2002.

[12] P. Sundararajan, O. J. Mengshoel, and T. Selker, "Multi-fisheye for interactive visualization of large graphs," in *Proceedings of the 17th AAAI Conference on Scalable Integration of Analytics and Visualization*, ser. AAAIWS'11-17. AAAI Press, 2011, pp. 46–50.

[13] V. Batagelj, A. Mrvar, and M. Zaveršnik, "Partitioning approach to visualization of large graphs," in *International Symposium on Graph Drawing*. Springer, 1999, pp. 90–97.

[14] L. C. Freeman, "Centrality in social networks conceptual clarification," *Social networks*, vol. 1, no. 3, pp. 215–239, 1978.

[15] S. P. Borgatti, "Centrality and network flow," *Social networks*, vol. 27, no. 1, pp. 55–71, 2005.

[16] E. M. Daly and M. Haahr, "Social network analysis for routing in disconnected delay-tolerant manets," in *Proceedings of the 8th ACM international symposium on Mobile ad hoc networking and computing*. ACM, 2007, pp. 32–40.

[17] S. Brin and L. Page, "Reprint of: The anatomy of a large-scale hypertextual web search engine," *Computer networks*, vol. 56, no. 18, pp. 3825–3833, 2012.

[18] "Pagerank implementation description," http://www.ccs.northeastern.edu/home/daikeshi/notes/PageRank.pdf, accessed: 2016-11-27.

[19] M. Brautbar and M. Kearns, "A clustering coefficient network formation game," in *Intl. Symp. on Algorithmic Game Theory*. Springer, 2011, pp. 224–235.

[20] R. Zafarani, M. A. Abbasi, and H. Liu, *Social media mining: an introduction*. Cambridge University Press, 2014.

[21] A. K. Jain and R. C. Dubes, *Algorithms for clustering data*. Prentice-Hall, Inc., 1988.

[22] V. D. Blondel, J.-L. Guillaume, R. Lambiotte, and E. Lefebvre, "Fast unfolding of communities in large networks," *Journal of statistical mechanics: theory and experiment*, vol. 2008, no. 10, p. P10008, 2008.

[23] M. E. J. Newman, "Modularity and community structure in networks." *Proceedings of the National Academy of Sciences of the United States of America*, vol. 103, no. 23, pp. 8577–8582, 2006.

[24] L. Waltman and N. J. van Eck, "A smart local moving algorithm for large-scale modularity-based community detection," *The European Physical Journal B*, vol. 86, no. 11, pp. 1–14, 2013.

[25] R. Rotta and A. Noack, "Multilevel local search algorithms for modularity clustering," *Jrnl. of Exper. Algorithmics*, vol. 16, pp. 2–3, 2011.

[26] K. Alsabti, S. Ranka, and V. Singh, "An efficient k-means clustering algorithm," 1997.

[27] "Human protein (vidal) network dataset – KONECT," Oct. 2016. [Online]. Available: http://konect.uni-koblenz.de/networks/maayan-vidal

[28] "Protein network dataset – KONECT," Oct. 2016. [Online]. Available: http://konect.uni-koblenz.de/networks/moreno_propro

Business Process Models as a Foundation for High-Fidelity-Prototypes of Mobile Enterprise Applications

Tobias Braumann, David Broll, Matthias Jurisch, Bodo Igler
Faculty of Design – Computer Science – Media
RheinMain University of Applied Sciences
Wiesbaden, Germany
Email: {tobiasbraumann, lustra87}@gmail.com, {matthias.jurisch, bodo.igler}@hs-rm.de

***Abstract*—Mobile Enterprise Applications are becoming more and more relevant to enterprises as the dissemination of smartphones has risen over the last decade. However, developing these applications is a very challenging and resource-intensive task. In this context, prototyping can lead to several benefits, including a better app usability. While Mobile Enterprise Applications are often used to support or carry out business processes, no prototyping approach exists that is based on business process models. In this paper, we present a tool that fills this gap. The tool uses a Business Process Model and Notation (BPMN) model annotated with screen designs as a source for generating a prototype. The prototype is integrated with a business process execution engine that runs the business process.**

Keywords–Mobile Enterprise Application; BPMN; Process Model; Prototyping.

I. INTRODUCTION

Since the beginning of the decade, mobile applications have become more and more ubiquitous. This trend also reached enterprises [1], where employees expect to use smartphone apps for their daily work with the high usability they are accustomed to from using consumer apps [2]. These expectations and the continuously and fast changing ecosystem of mobile app development pose a significant difficulty for the development of Mobile Enterprise Applications (MEA).

MEAs differ from regular consumer apps in several ways. E.g., they are often used to support some kind of business process, have only few potential users in comparison to consumer apps and need to adhere to enterprise specific guidelines [2]. Integrating business processes into mobile applications requires implementing new interfaces to process engines and adhering to process guidelines. These are some of the challenges for MEA development that are caused by MEA-specific aspects. Since other aspects of mobile application development also need to be taken care of, these factors contribute to developing MEAs being a time consuming and expensive process.

To reduce the effort required to develop mobile applications in general, prototyping can be used. A good prototyping process can prevent misunderstandings and make the conceptual phase of the development process significantly easier to handle and therefore reduce costs [3][4]. More importantly, this can also lead to a better usability of the final product. This will improve the willingness of employees to use the final MEA. However, no prototyping tool that supports all of the aforementioned aspects of MEAs exists. To our knowledge, there is no prototyping approach that caters to the business process aspect of this type of application.

In this paper, we present a prototyping approach that focuses on using business process models [5] as the primary source for MEA prototypes. This is accomplished by annotating process models with screen designs and creating a prototype using code generation and business process execution engines that can interpret and automatically execute business processes. This work is embedded into the scope of the Prototyping Framework for Mobile App Design in Large Enterprises (PROFRAME) [6]. The presented work will lay the foundation for the implementation of PROFRAME.

The remainder of this paper is structured as follows: Section II gives a brief overview on related work and identifies the research gap. The general approach of this paper is presented in Section III. Details on the behavioural modeling of the prototypes are given in Section III-A, designing screens is discussed in Section III-B and code generation and prototype execution are presented in Section III-C. The implementation is described in Section III-D. Section IV discusses advantages and disadvantages of our approach. A conclusion is given in Section V.

II. RELATED WORK

According to [2], a huge gap between the development of MEAs and standard non-mobile enterprise applications can be observed. However, the demand for MEA development in the next few years will be much higher than the supply [1]. Hence, it is important to support a very efficient way of implementing MEAs.

One way to improve the development of MEAs is improving the prototyping process. Several models for classifying prototypes have been proposed in the literature. Nielsen [4] proposed a distinction between vertical and horizontal prototyping fidelity. A horizontal prototype supports most functionalities of a product, whereas a vertical prototype supports only a few functionalities but is technically more similar to the final product. The filter fidelity model [7] adds more dimensions to this view, e.g., regarding interactivity, data model, weight and many other dimensions. Breadth and depth of functionality are also included in this model.

For prototyping mobile applications in general, many products and approaches can be found in the literature. However, regarding prototyping for MEAs, only a few tools can be found (e.g., Kony, Verivo Akula and SAP Mobile) [8]. These tools are often focussed on a specific use case or bound to a specific platform. None of them take business process modeling into account, so the depth of functionality according to the filter fidelity model is low.

Integrating process models into application development has been discussed in the area of process-driven development. AgilePDD [9] proposes an agile approach to implementing business processes. In the prototyping phase of this process, business process models are used to define the behaviour of the prototype. While this approach seems promising, it does not define how a prototype should be generated from the process model or how process steps should be represented as screens. The approach is in general focussed on modeling use cases with business process models, whereas generating code from these models is only mentioned as an option to be considered [10].

From the presented literature, we can conclude that no prototyping approach or tool for MEAs exists that facilitate a high fidelity regarding the representation and integration of business process models into the prototype. This issue is at the core of our research, since a prototype that better resembles the final product improves its usability.

III. Approach

We consider three major requirements for our work: the tool needs to support (1) modeling a business process, (2) designing a user interface and (3) generating a platform-independent prototype that can be executed on mobile devices. The basic idea of our approach is to use business process models as the primary source for the prototype. The process model defines the behaviour of the app. To add a graphical user interface, the process model is annotated with screen designs for specific parts of the business process. With this information, a prototype for the app is generated.

The Business Process Model and Notation (BPMN) [11] is the most popular language for modeling business processes [12]. In practice, several implementations of process engines exist. They are able to interpret and execute BPMN models and integrate business processes with several backend systems. Therefore, BPMN is used as the process model language in the presented prototyping tool.

An overview of the prototyping process is given in Figure 1. A Business Analyst/UX-Designer uses the *BPMN/Form Modeler* to create a model of the process that shall be implemented with an app including a user interface design. To model the process itself, one can simply use an existing process modeling tool that supports BPMN. This can be done in close coordination with the customer, e.g., at a kick-off-meeting for a project. The result of this process is an *extended BPMN file*. This file can then be used to generate a *Web App* that cooperates with a business *process engine*. The customer can then use this app as a prototype, which allows a clear separation of the code generation and the prototype modeling. For customization, a developer that modifies the code generation can be included in the process.

To support the described process, answers to the following questions are required: What aspects of a process should be represented as screens (Section III-A)? How can screens be designed and how can data be reused over several screens (Section III-B)? How are prototypes generated (Section III-C)?

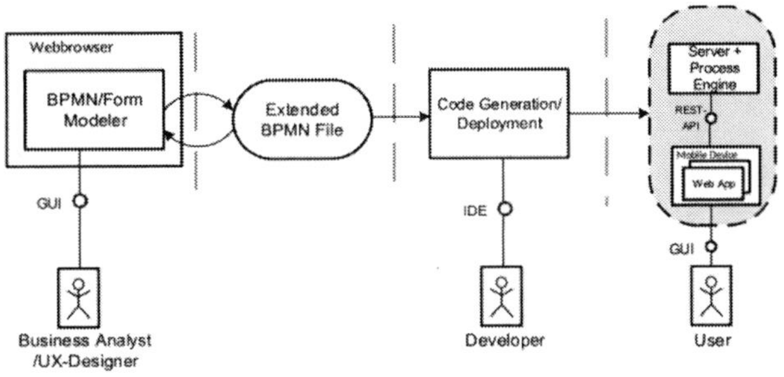

Figure 1. Architecture of the approach

A. *Process Model*

BPMN in general is well-known for its graphical representation of business processes. An example model is shown in Figure 2. The most important element of BPMN is the *task* (e.g., *Apply for Vacation*). Tasks represent any kind of activity. Several kinds of tasks exist, the kind of task is represented by an icon at the top of a task. *Apply for vacation* is a user task and *Check Vacation request* is a script task. User tasks require user interaction whereas script tasks are automatically executed by the business process engine.

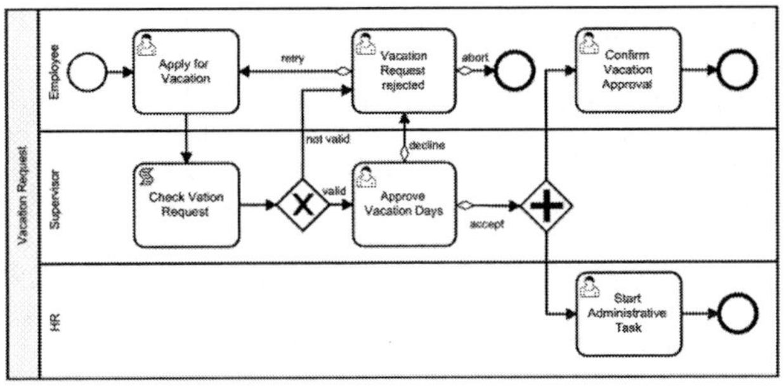

Figure 2. Example process

To connect tasks, so called *Sequence Flows* that are represented by arrows are used. Gateways (represented by rhombuses) are used to model situations where the flow is split, either because of decisions (x) or parallel execution (+). The swimming pool element (*Vacation Request*) is used to structure the control flow. A swimming pool can contain multiple swimlanes (e.g., *Employee*) that distinguish different domains of activity.

Our approach proposes a representation of tasks as screens: when the model is executed, each user task corresponds to one screen on the mobile device. A swimming pool corresponds to an app and a swimlane corresponds to a user role. For the example shown in Figure 2, users with role *Employee* would be shown at most three different screens (*Apply for Vacation, Vacation Request Rejected,* and *Confirm Vacation Approval*) and users with role *Supervisor* or *HR* one (*Approve Vacation Days* and *Start administrative Task*). Sequence flows determine the control flow of the business process.

B. *Form Modeler*

To design the forms that correspond to user tasks, a form modeler is used that is able to store screen designs as

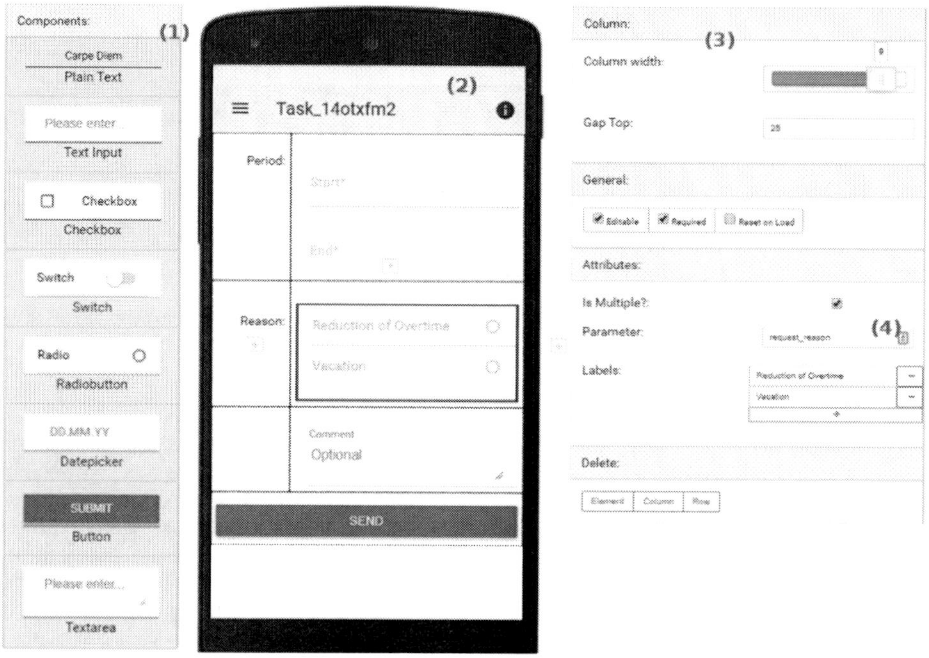

Figure 3. Form Modeler Screenshot

annotations in BPMN files. The form modeler needs to add a screen design to each user task and store the design as an annotation in the BPMN model. A screenshot of the form modeler that implements this idea is shown in Figure 3. The user of the form modeler can drag and drop user interface components (1), e.g., *Plain Text*, *Text Inputs* and *Radio Buttons*, into the screen layout (2). Properties of components can be modified using controls on the right (3).

Our approach uses a grid layout to model the screen design. By using a grid layout, the prototype is not bound to a specific screen size or orientation. The grid is shown as dashed lines in the screenshot. Users can add and remove rows and columns. Each cell in this grid can only hold one widget. To improve the design, the user can modify row height and column width.

By modifying a component's properties using the box on the right (3), the user can edit several aspects regarding its behaviour and appearance. E.g., inputs can be set as editable and required and their label can be defined. The property *parameter* (4) is used to specify parameter ids that are used to identify data throughout the business process. The parameter ids are identifiers in a global data space bound to a workflow. When a screen is used to input data into a field with a certain parameter id and another screen shown later in the process has a component with a matching parameter id, the second screen will show the data entered in the first screen.

The screen shown in the example corresponds to the task *Apply for Vacation* from Figure 2. To view the data entered into this screen, e.g., in the task *Approve Vacation Days*, it is only required to add a UI component to that task and set its parameter to `request_reason`, similar to the example shown in Figure 3 (4).

C. Code Generation and Process Execution

The previously described steps allow the creation of an annotated BPMN model that contains information about the behaviour of the application, as well as the UI design. Based on this information, code generation can be used to create a prototype.

Besides allowing generating app prototypes, using BPMN as a foundation for the prototype allows execution of the process model on a business process execution engine. To exploit this circumstance, the generated prototype is separated into two parts: (1) A business process engine that is given the business process model and executes it and (2) a prototype that interacts with the business process execution engine. The engine controls the process and data related to it. This allows the synchronization between prototypes for different user roles involved in the process, which are all created in the generation process. Also, the business process engine can be integrated with other enterprise systems, which allows accessing real-world data by the prototype.

D. Implementation

As a component for modeling business processes, Camunda Modeler [13] is used. Camunda Modeler needed to be extended to provide an interface to the form modeler. Angular [14] is used to implement the form modeler from scratch to allow a seamless integration with the process modeler. These components write their data into the shared extended BPMN file.

Prototype generation form the shared BPMN documend is implemented using XSLT [15]. To support multiple mobile platforms, the generated code uses the Ionic framework [16], a HTML and Javascript-based Framework, as SDK. With Ionic, the visual design of the app can be easily changed using CSS. This supports the integration of enterprise corporate design guidelines into the product. To execute the business process, the Camunda Core Engine [17] is used. To interact with the engine, the prototype uses the Camunda REST API.

IV. Discussion

Designing a prototype for the business process shown in Figure 2 using the presented method takes about 25 minutes. After presenting the application example to experts from the MEA development field, they concluded that it would take one or two days for a developer to create a similar app. Since a developer needs to be involved in this process in addition to a business analyst, some communication overhead is also very likely. This indicates that the approach is able to create prototypes with a reasonable effort and is therefore able to compete with other prototyping approaches.

In comparison to prototyping approaches mentioned in Section II, we see several benefits. One important advantage of using BPMN as the foundation for prototyping is that it supports reusing existing process models to create a prototype. Even an automated transformation from existing files is possible. This is not possible for prototyping approaches based on other models. Another important aspect of this approach is the possibility to build applications using more than one user role easily. Supporting a business process execution engine allows the integration of existing enterprise systems into the prototype, since these systems can be integrated into the process engine. This allows the prototype to access real-world data, which give the user of the prototype a better understanding of the functionality.

A drawback of our approach is the limitation regarding visual design choices of the form modeler caused by the grid layout and the limited set of standard components. While inexperienced users might see the simplicity as an advantage, especially designers might need more freedom in positioning components and a broader collection of usable widgets.

Regarding the prototype's fidelity according to the filter fidelity model, our approach allows building prototypes that have a high breadth and depth of functionality and a close relation to data and appearance of the final product. This makes it easier to demonstrate to customers how an app can support their business processes and help manage expectations. This can lead to reduced costs for reworking requirements and app concepts during the MEA development process and improve the usability of the final product.

V. Conclusion and Future Work

In this paper, we have presented a prototyping approach for MEAs that is based on the usage of business process models written in BPMN. Prototypes are created using a business process annotated with screen designs. The annotated process is then used to generate a prototype that consists of an app and a business process execution engine that executes the process.

This approach facilitates fast prototyping of MEAs, since it is possible to reuse existing BPMN models for prototyping and integration with other enterprise applications through the business process execution engine. The generated prototypes can achieve a high level of fidelity regarding several aspects. Especially the depth of functionality and the visual quality of the prototype is high.

As future work, we plan to quantitatively evaluate the benefits of this prototyping approach regarding the ability to develop MEAs. Another next step could be the integration of existing screens from a standard screen library into the prototyping tool.

Acknowledgment

This work was funded by the German Federal Ministry of Education and Research, grant no. 03FH032PX5; the PROFRAME project at RheinMain University of Applied Sciences. All responsibility for the content of this paper lies with the authors.

References

[1] Gartner, "Gartner says demand for enterprise mobile apps will outstrip available development capacity five to one," https://www.gartner.com/newsroom/id/3076817, website [retrieved: September, 2017], 2015.

[2] A. Giessmann, K. Stanoevska-Slabeva, and B. de Visser, "Mobile enterprise applications–current state and future directions," in 2012 45th Hawaii International Conference on System Science (HICSS), Jan 2012, pp. 1363–1372.

[3] F. Kordon and Luqi, "An introduction to rapid system prototyping," IEEE Transactions on Software Engineering, vol. 28, 2002, pp. 817–821.

[4] J. Nielsen, Usability Engineering. San Francisco, CA, USA: Morgan Kaufmann Publishers Inc., 1993.

[5] R. S. Aguilar-Saven, "Business process modelling: Review and framework," International Journal of Production Economics, vol. 90, no. 2, 2004, pp. 129–149.

[6] M. Jurisch, B. Igler, and S. Böhm, "PROFRAME: A Prototyping Framework for Mobile Enterprise Applications," in CENTRIC 2016, The Ninth International Conference on Advances in Human-oriented and Personalized Mechanisms, Technologies, and Services, 2016, pp. 7–10.

[7] T. Hochreuter, K. Kohler, and M. Mareen, "Towards a more conscious use of prototypes in mobile user experience design," in Prototyping to Support the Interaction Designing in Mobile Application Development (PID-MAD), 2013.

[8] J. Wong, V. L. Baker, A. Leow, and M. Resnick, "Magic quadrant for mobile app development platforms," https://www.gartner.com/doc/3744122/magic-quadrant-mobile-app-development, website. [retrieved: September, 2017], 2015.

[9] A. Herden, P. Farias, P. de Andrade, and A. Albuquerque, "Agile PDD: One approach to software development using BPMN," in 11th International Conference Applied Computing, Porto, Portugal, 2014.

[10] A. Herden, P. Farias, and A. Albuquerque, "An approach based on BPMN to detail use cases," in New Trends in Networking, Computing, E-learning, Systems Sciences, and Engineering. Springer, 2015, pp. 537–544.

[11] Object Management Group, "Business process model and notation, version 2.0," http://www.omg.org/spec/BPMN/2.0/PDF/, website. [retrieved: September, 2017], 2011.

[12] P. Harmon, "The state of business process management 2016," http://www.bptrends.com/bpt/wp-content/uploads/2015-BPT-Survey-Report.pdf, technical report. [retrieved: September, 2017], 2016.

[13] Camunda, "Modeler - Camunda BPM," https://camunda.org/features/modeler/, website. [retrieved: September, 2017], 2017.

[14] Google, "Angular," https://angular.io, website. [retrieved: September, 2017], 2017.

[15] W3C, "XSL Transformations (XSLT) Version 1.0," https://www.w3.org/TR/xslt, website. [retrieved: September, 2017], 1999.

[16] Drifty Co., "Build Amazing Native Apps and Progressive Web Apps with Ionic Framework and Angular," https://ionicframework.com/, website. [retrieved: September, 2017], 2017.

[17] Camunda, "Engine - Camunda BPM," https://camunda.org/features/engine/, website. [retrieved: September, 2017], 2017.

The Attitude of German Web Users Toward Foreign Websites for Buying Products Online

Bastian Eine, Werner Quint
Faculty of Design – Computer Science – Media
RheinMain University of Applied Sciences
Wiesbaden, Germany
email: bastian.eine@hs-rm.de, werner.quint@hs-rm.de

***Abstract*—Today, consumers tend to buy products online because of several advantages compared to shop in brick and mortar stores. One advantage of online shopping is to browse, search, and buy products from online shops all over the world. However, the attitude of online shoppers towards using foreign websites for buying products online might be influenced by factors that result from high uncertainties when using foreign websites. This paper analyses factors that might influence the attitude of web users toward foreign websites to buy products online. In this study, a survey with 60 German web users was conducted. The results were presented in a regression model and suggest that future research should focus on developing, testing and validating concepts that are influencing the attitude toward foreign websites before conducting large-scale studies.**

Keywords-online shopping; foreign websites.

I. INTRODUCTION

Advantages of online shopping versus shopping in brick and mortar stores are convenience and reducing cost for information. Web users can easily screen information of a large variety of products and compare products and their prices [1]. Online shops are available anytime from anywhere. Buying products online is time efficiently because of avoiding crowds in brick and mortar stores and the traveling to get there [2]. Also, discrete shopping, avoidance of pressure from salesmen [3], and personalized shopping through recommender systems are additional reasons for shopping online for some users [4]. Furthermore, the geographical limitation of brick and mortar stores is overcome by online shopping. Users can browse, search, and buy products from online shops all over the world. Literature has shown that there are several factors that influence web users' choice when selecting an online shop to buy products from. However, there are only few studies available that explicitly examine factors that might have an influence on the attitude of online shoppers toward using foreign websites for buying products online [5][6].

While several studies analyzed customer's attitudes and factors influencing online shopping behavior in general [7][8][9], factors like risk, trust, privacy and security issues might have a negative influence on customer's attitude toward using online shops of foreign countries, because of high uncertainties (e.g., country's laws or reliability of shipping) [10]. On the other hand, there might be factors influencing the customer's attitude positively toward foreign online shops, e.g., originality of the product or lower prices. In addition, there might be some disadvantages of shopping online in general that can become even more relevant when shopping on foreign websites. For example, higher costs for shipping and handling charges, or the time it may take until the product arrives at the consumer may exceed the advantages of shopping online for consumers [11].

This study aims to analyze factors that influence the attitude of web users toward foreign websites for buying products online. It is important to understand customers' characteristics, online shopping behavior, consumer' expectations, and their attitude toward online shopping in general to get an insight on their attitude toward foreign online shops. In this study, a conceptual model and hypotheses were developed and tested by conducting a web-based questionnaire on German web users and their attitude toward foreign websites for buying products online. This research will contribute to the understanding of the adoption of online shopping and will support the development of international online shopping in general. In addition, this study can encourage more researchers to examine factors that improve the attitude of web users of different countries toward foreign websites for buying products online.

After this introduction, a review of the literature about influences on the attitude towards online shopping and the hypotheses for this study will be described in section 2. In section 3, the methods for the survey that was conducted will be explained. Section 4 will present the results of this survey. Section 5 will summarize the main findings of this study and will give an outlook on further research.

II. LITERATURE REVIEW

A. *Attitude Toward Websites for Buying Products Online*

The high popularity of online shopping is still unbroken. According to estimates by private consultancies [12][13][14], worldwide retail e-commerce sales is forecast to double from 1.9 trillion US-Dollars in 2016 to 4.1 trillion US-Dollars in 2020 [13]. Although, the forecast is that e-commerce growth will slow over time, a 22% growth rate in 2018 still indicates the importance of e-commerce channel for consumers and retailers. The number of web users buying products online is also still growing and had been expected to be over 1.6 billion people by the end of 2017 [13].

Several research had been conducted about what might influence consumers in their choice of browsing and buying products online. Studies in the context of Theory of Planned Behavior (TPB) and Technology Acceptance Model (TAM) have shown that there is a strong positive impact of customers' attitude toward online shopping on the customers' intention to use online shopping (e.g., [15][16]). The attitude toward online shopping can be defined as the "the extent to which a consumer likes on-line shopping and considers it to be a good idea" [9].

Because of the large quantity of research about online shopping acceptance in general, some researchers already tried to summarize the existing findings in comprehensive models. Chang, Cheung, and Lai [17] analyzed empirical studies where factors for adoption of online shopping were analyzed. They used and adapted the scheme by Jarvenpaa and Todd [18] to categorize these factors into three categories. These categories are perceived characteristics of the web as a sale channel, characteristics of the consumer, and characteristics of the website or products.

Perceived characteristics of the web as a sales channel describe uncertainty and concerns regarding trust, risk, privacy and security. These risks can relate to the product (quality, expectation) or the transaction process (payment, delivery). In addition, the perceived relative advantage (convenience, time saving) of online shopping compared to brick and mortar stores, the online shopping experience of the web user and the service quality can play an important role when evaluating the web as a sales channel.

The second category is the characteristics of the consumer. These are the consumer's shopping orientation (i.e., price, recreational, brand, and impulsiveness), demographics, knowledge of and experience with the Internet, as well as the consumer's innovativeness in general.

The third characteristics relate to the website or product, which can reduce (providing additional service, e.g., money back guarantee) or raise the perceived risk (e.g., high cost product, infrequently purchased).

Similar work had been done by Li and Zhang [19] before, who found less but similar factors and categories. Zhou, Dai, and Zhang [20] also combined research about consumer characteristics that might influence the attitude toward online shopping to develop an Online Shopping Acceptance Model. Their model incorporated factors about demographics, Internet experience, shopping motivation, innovativeness, perceived outcome, shopping orientation, normative beliefs attitude, online shopping experience, online shopping intention and satisfaction.

Because of the potential of online shopping in international markets, several researchers have also analyzed the influence of culture on the adaption of online shopping [21][22][24][25]. Most studies applied Hofstede's cultural dimensions to explain differences in national culture and online shopping. Research has shown that success of a company's online shop in one country does not have to imply success in other countries or cultures [26][27] because customers' characteristics and customers' expectations of online shopping and product information may be different from one country to another [28].

B. Attitude Toward Using Foreign Online Shops

Although there has been several research on cultural dimensions and the adaption of online shopping to national culture or countries, there are only few studies available that examine consumers' attitude toward using foreign websites in general. Hence, research demands for more empirical studies on "factors affecting global online consumers' willingness to transact on international web sites" [29].

Beside culture, factors regarding economy, infrastructure, politics and laws of other countries can have an influence on a global consumers' attitude toward online shopping on foreign websites [30]. Furthermore, consumers' intention to buy products from foreign websites can be influenced by consumers' characteristics (online shopping confidence and expertise) and motivations (importance of price, product availability in domestic market, time of delivery, and shipping costs), as well as the consumers' knowledge about other countries [5]. In addition, the frequency of using foreign websites for online shopping can be different from country to country. For example, Australians tend to order many of their online purchases from foreign websites [5]. Many Americans have bought products from foreign websites once, but might be detained by high costs for transaction and shipping to buy more frequently overseas. In contrast, only 2% of Swedish consumers have bought online from foreign websites [31].

Attitude toward using foreign online shops can be defined as the extent to which a consumer likes online shopping on foreign websites and considers it to be a good idea (modified by Vijayasarathy [9]). In this study, a foreign online shopping website is defined as an online shopping website where it is obvious to the customer that the product is shipped from abroad, that seller has a foreign address, or that the website's address has an obviously foreign ending.

This study analyzes consumer characteristics that have had an influence on attitude in other studies and that are hypothesized to be especially relevant in the context of foreign websites for buying products online. These factors are financial risk, product risk, convenience risk, non-delivery risk, return policy risk, personal innovativeness, and relative advantage of online shopping in general.

1) Risks

Risk is consumers' uncertainty that they face when they cannot foresee the consequences of a particular transaction [39]. Thus, risk in online shopping can be defined as "the potential for loss in pursuing a desired outcome while engaged in online shopping" [33]. Risk and trust have been an object of investigation in several studies because of the impersonal characteristics of online shopping compared to shopping in brick and mortar stores [34]. Research had shown that the influence of perceived risk on customer's intentions to buy online was strongly negative [16][35][36]. Most research has also analyzed different types of risks in online shopping. For example, product uncertainty and transaction risk have proven to be a problem in B2C e-commerce [37]. Risk had proven to have a negative influence on online shopping intention across several countries and cultures [22][28][38]. In this study, the types of risks in online shopping (financial, product, convenience, non-delivery, and return-policy) adapted from [39] will be analyzed and adapted to German web users and the influence of risk on web user's attitude toward foreign websites for buying products online. Financial risk describes risks regarding payment, product risk describes risks regarding product viewed on screen, convenience risk describes risks regarding comfort of online shopping, non-delivery risk describes risk regarding shipping, and return-policy risk describes risk regarding sending products back to the retailer. Also, in other studies financial risk, product risk, concern for privacy and security have shown to have significant influence on attitude [40]. Thus, hypotheses regarding risks are:

H1a: Perceived financial risk in general has a negative influence on attitude toward foreign websites for buying products online.

H1b: Perceived product risk in general has a negative influence on attitude toward foreign websites for buying products online.

H1c: Perceived convenience risk in general has a negative influence on attitude toward foreign websites for buying products online.

H1d: Perceived delivery risk in general has a negative influence on attitude toward foreign websites for buying products online.

H1e: Perceived return policy risk in general has a negative influence on attitude toward foreign websites for buying products online.

2) Personal Innovativeness

Personal innovativeness can be defined as the extend and speed an individual adopts new innovations [41]. In comparison to shopping in brick and mortar shops, shopping online can be characterized as innovative. Several studies analyzed personal innovativeness or domain specific innovativeness and the influence on attitude or intention to use online shopping, also for shopping across national borders [6] and cultures [42]. Most studies have shown that personal innovativeness has a positive influence on attitude and intentions to shop online [15][43][44]. Nevertheless, it has also shown that domain specific innovativeness (product specific innovativeness in online shopping context) can lead to more significant findings for this relationship [45]. However, this study does not explore attitude toward online shopping regarding a specific product or product type. Hence, this study uses personal innovativeness as the concept that influences attitude toward online shopping on foreign websites.

H2: Personal innovativeness regarding information technology has a positive influence on attitude toward foreign websites for buying products online.

3) Relative Advantage

Relative advantage is defined as the extent to which a consumer perceives an innovation to be better than the idea it supersedes [41]. The advantages of online shopping in comparison to shopping in traditional physical stores can be lower prices, saving time, large variety of products, ease of comparing products, 24 hours availability of online shops and convenience (e.g., buying from anywhere, avoiding queues) [2]. Research has shown that relative advantage has a positive influence on attitude toward online shopping in general [8], also in international studies [23].

H3: Perceived relative advantage in general has a positive influence on attitude toward foreign websites for buying products online.

III. METHODS

A. *Sample and Data Collection*

This research focuses on the sample of web users from Germany. Convenience sampling was used for sample selection. The questionnaire was created with Google Forms and distributed via QR-code and online via e-mail or WhatsApp message by sending a hyperlink. The questionnaire was filled out online only. In total, 60 questionnaires were completely filled. Demographic information and information on web users' internet as well as online shopping experience are summarized in Tab. 1.

TABLE I. DESCRIPTIVE STATISTICS OF THE SAMPLE

Demographic Factor	**Descriptive Statistic**
Age (in year)	19-21: 13 (21.7%) 22-29: 24 (40.0%) 30-39: 13 (21.7%) 40-49: 5 (8.3%) 50 and older: 5 (8.3%)
Gender	Male: 32 (53.3%) Female: 28 (46.7%)
Education	Below Bachelor's degree: 42 (70.0%) Bachelor's degree: 6 (10.0%) Master's degree: 12 (20.0%)
For how long have you been using the Internet (in years)?	4-6: 7 (11.7%) 7-10: 16 (26.7%) More than 10: 37 (61.7%)
How many hours per week do you spend using the Internet?	1-3: 7 (11.7%) 4-6: 3 (5.0%) 7-10: 12 (20.0%) More than 10: 38 (63.3%)
For how long have you been using the Internet for shopping (in years)?	1-3: 14 (23.3%) 4-6: 20 (33.3%) 7-10: 20 (33.3%) More than 10: 6 (10.0%)
How often do you buy products online?	1-2 times a year: 3 (5.0%) 3-4 times a year: 9 (15.0%)) Once every 1-2 months: 32 (53.3%) 2-3 times a month and more: 16 (26.7%)
For how long have you been using the Internet for shopping on foreign websites?	Never: 15 (25.0%) Less than 1 year: 7 (11.7%) 1-3 years: 19 (31.7%) 4-6 years: 10 (16.7%) 7-10 years: 8 (13.3%) More than 10 years: 1 (1.7%)
How often do you buy products from foreign websites?	Never: 15 (25.0%) Less than once a year: 23 (38.3%) 1-2 times a year: 12 (20.0%) 3-4 times a year: 7 (11.7%)) Once every 1-2 months: 3 (5.0%)

B. Measures

The main dependent variable attitude toward foreign websites for buying products online was measured using the modified scale by Vijayasarathy, Limayem, Khalifa and Frini [9][15] adapted from TBA [55]. Sample items for the dependent and independent variables are listed in Tab. 2.

TABLE II. LIST OF ITEMS FROM QUESTIONNAIRE

Construct	Question Wording	Source
Attitude toward foreign websites for buying products online	Online shopping on foreign websites is a good idea (ATF1) Purchasing through the Web on foreign websites is convenient (ATF2) I don't like buying products online on foreign websites (reversed) (ATF3) Buying products on foreign websites is attractive (ATF4)	[9][15]
Financial risk	I feel that payment data is safe (reversed) (FIN1) I might get overcharged as the retailer has my payment data (FIN2) I feel that my personal information given for transaction to the retailer may be compromised to 3rd party (FIN3)	[39][48] [49]
Product risk	I always get what I ordered (reversed) (PRO1) I might receive malfunctioning merchandise (PRO2) It is hard to judge the quality of merchandise over the internet ((PRO3))	[39][48] [49]
Convenience risk	Finding right product is difficult (CON1) I cannot wait till the product arrives* (CON2) I cannot get to examine the product before buying (CON3) I feel that it will be difficult settling disputes with the retailer (CON4) It is easy to cancel orders (reversed) (CON5) I will have problem in returning product bought online (will have to send the product back through some shipper and wait to see if the retailer accepts it without any hassle) (CON6)	[39][48] [49]
Non-delivery risk	I might not receive the product ordered online (DEL1) Only if there is a reliable shipper* (DEL2) The ordered product might not be shipped (new) (DEL3)	[39][48]
Return-policy risk	There does not have to be free return shipment service available (reversed) (RET1) Only if I can return the product without any frills or strings attached (RET2) Only if there is a money back guarantee (RET3)	[39]
Personal innovativeness	If I heard about a new information technology, I would look for ways to experiment (PIN1) Among my peers, I am usually the first to try out new information technologies (PIN2) In general, I am hesitant to try our new information technologies (reversed) (PIN3) I like to experiment with new information technologies (PIN4)	[43][50]
Relative advantage	The Internet is the best place to buy items that are hard to find (REL1) Shopping online is convenient (REL2) Shopping online does not save time (reversed)* (REL3) The Internet is the best place to find bargains (REL4)	[23]

* Marked items were not used in correlation and regression analysis of this work.

The scale was also adapted in several studies to measure the attitude toward technology in TAM [47]. The items were measured using Likert scale ranging from 1 (strongly disagree) to 5 (strongly agree).

Demographic factors and factors related to Internet and online shopping experience were included in the model as control variables. These factors are age, gender, education. Age was measured in years, gender was measured as a dummy variable (male was coded 1, female was coded 0), education, Internet experience and online shopping experience were measured ordinally.

C. Data Analysis

Data analysis was performed using Multiple Ordinary Least Squares (OLS) regression. Multiple OLS regression is a statistical technique that provides information about the relationship between several independent variables and a dependent or criterion variable. The analysis was performed using IBM SPSS Statistics version 22.

IV. RESULTS

Before performing the regression analysis, a few steps need to be conducted. First, construct reliability of the main concepts that are measured using multiple item scale were evaluated by using Cronbach alphas coefficient. Three items from the questionnaire were removed (CON2, DEL2, ADV3) to improve Cronbach alpha coefficients. Three of the concepts exceed the widely suggested value of 0.7 [51], which means that Cronbach alpha coefficient of convenience is acceptable (>0.7) and Cronbach alpha coefficient of personal innovativeness and attitude toward foreign websites for buying products online are good (>0.8). For these constructs, the reliability is satisfactory. The Cronbach alpha coefficient for financial risk only exceeds the value of 0.5. This indicates that the reliability of this construct is poor. The Cronbach alpha coefficient for delivery risk and return policy risk only exceeds the value of 0.6. This indicates that the reliability of this construct is questionable. The Cronbach alpha coefficient for product risk does not exceed the value of 0.5. This indicates that the reliability of this construct is not acceptable. The results are shown in Tab. 3. Then, the summated scale for each concept was created by averaging the multiple-item scales that belong to the same concept to perform regression analysis.

TABLE III. RESULTS FROM RELIABILITY TEST

Variables	*Cronbach alphas (α) coefficient*
Financial risk	0.501
Product risk	0.461

Variables	*Cronbach alphas (α) coefficient*
Convenience risk	0.747
Delivery risk	0.626
Return policy risk	0.659
Personal innovativeness	0.819
Relative advantage	0.619
Attitude toward foreign	0.888

Finally, Pearson correlation coefficients were used to analyze bivariate correlations among variables. Thus, one-on-one relationships between key variables were examined by using correlation analysis. Results from correlation analysis are presented in Tab. 5.

The results show that online shopping experience in years is higher for male (r= 0.330; p<0.01) and for older respondents (r= 0.335; p<0.01), but lower for respondents with a high return policy risk (r= -0.444; p<0.01) and respondents with low personal innovativeness (r= 0.364; p<0.01). Online shopping frequency is higher for respondents that have higher personal innovativeness (r= 0.367; p<0.01) and respondents that have higher relative advantage (r= 0.439; p<0.01). Respondents that have a higher convenience risk tend to have a lower personal innovativeness (r= 0.521; p<0.01) and lower relative advantage (r= -0.307; p<0.05).

The hypotheses in this study were tested by multiple OLS regression analysis. The results reveal that the influence of product risk (β=-0.108; p=0.444), convenience risk (β=-0.310; p=0.063), and return policy risk (β=-0.136; p=0.358) on attitude toward foreign websites for buying products online is negative, but not statistically significant (p-values >0.05). Financial risk has a positive influence on attitude toward foreign websites for buying products online (β=0.207; p=0.162) but is also not significant (p>0.05). Therefore, hypotheses 1a, 1b, and 1d are not statistically supported. Delivery risk has a positive influence on attitude toward foreign websites for buying products online (β=0.321; p=0.028) and is significant (p<0.05). Hence, hypothesis 1c is not supported, too. Hypotheses 2 and 3 predict a positive influence of personal innovativeness and relative advantage on attitude toward foreign websites for buying products online. The results are reported in Tab. 4.

TABLE IV. REGRESSION RESULTS

Variables	Coefficients
Constant (attitude toward foreign websites for buying products online)	3.645*
Main independent variables	
Financial risk	0.207
Product risk	-0.108
Convenient risk	-0.310
Delivery risk	0.321*
Return policy risk	-0.136
Personal innovativeness	0.022
Relative advantage	0.089
Control variables	
Online shopping experience in years	-0.092
Online shopping frequency	0.034
Age	-0.306*
Gender (male=1)	0.127
Education	0.049
R square	0.323
Adjusted R square	0.150

Notes: * p <.05, ** p < .01.
Standardized coefficients are reported.

It shows that these two variables have a positive influence but the results are not statistically significant (personal innovativeness (β=.022; p=0.891), relative advantage (β=.089; p=0.566). Therefore, hypothesis 2 and 3 are not supported.

The relationships between control variables and attitude toward foreign websites for buying products online are found as the following. Age negatively influences attitude toward foreign websites for buying products online (β=0-.306; p<0.05). Influences of gender (β=0.127), education (β=0.049), online shopping experience in years (β=-.092), and online shopping frequency (β=0.034) on attitude toward foreign websites for buying products online are not significant in this study. The values of R square (R^2=0.323) and adjusted R square (adj. R^2=0.150) in the regression model suggest that there might be more factors that can explain variance in the attitude toward foreign website for buying products online.

Finally, to check for the possible problem of multicollinearity among all variables in each equation, the Variance Inflation Factor (VIF) statistics was evaluated. The VIF values range from 1.281 to 2.018, which were significantly below the critical value of 10 as suggested by Moon, Chadee, and Tikoo [52]. This implies no multicollinearity issue in the analysis.

V. DISCUSSION

The present study aims to investigate the attitude of German web user toward foreign websites for buying products online. No supporting evidence that perceived financial risk, product risk, convenience risk, and return policy risk associate with attitude toward foreign websites for buying products online can be found from regression analysis with the data collected in this study. Also, the data is not sufficient to support the hypotheses that personal innovativeness and relative advantage influence the attitude toward foreign websites for buying products online. Results for the relationship between non-delivery risk and attitude toward foreign websites for buying products online showed a positive influence which is contrary to the hypothesis (positive influence). However, data shows that customers who are of younger age tend to report higher attitude toward foreign websites for buying products online.

Despite this finding, there are several weaknesses that need to be discussed. First, the samples selected for this study were collected by using convenience sampling. Thus, questionnaire was only distributed to web users the author already had contact to before. Small-scale data collection and the use of OLS regression with 7 parameters may limit the potential for statistically significant results and generalizibility. Future studies may need to expand the sample size in order to get results that are more reliable. Second, there are some doubts that the concepts and items that were adapted from [39] can lead to good reliability of concepts, which is a major weakness of this study. Furthermore, there might be several other concepts that are more suitable to explain the attitude toward foreign websites for buying products online [17][20]. Also, factors that describe consumers' knowledge about other countries could have been incorporated in such studies. However, a study that tried to analyze global consumers' characteristics [5] did not mention the items or sources they used for measuring in their work. Thus, future research should develop, test and validate concepts to fill this gap. In addition, respective items for questionnaires are needed to avoid collecting data that may cause bias or inaccuracy in the measurement. Also, this study analyzed German web users only. More countries can be analyzed to determine their attitude toward foreign websites. Thus, similarities and differences can be examined. Furthermore, perceived risk and attitude toward online shopping might be influenced by other factors [17][20][26]. Research has shown that trust, especially in international context, plays an important role on the perceived risk [22]. Moreover, risk is not only influenced by cultural characteristics of consumers, but also by perceived security of transaction systems or policies on payment system and privacy protection [42]. These might also differ from country to country. Hence, customers' attitude toward foreign websites might also depend on the foreign country, which they think of when they evaluate their attitude toward foreign websites. In addition, attitude toward buying a product online seems to be dependent on the type of product [43][53]. Altogether, it seems to be almost impossible to analyze all factors in one model. Still, because the geographical and cultural distance between customers and online retailers might be even higher foreign websites than domestic websites it can be expected that also perceived risk is higher for customers using foreign websites. Hence, future studies should explore this hypothesis by comparing attitude toward domestic with attitude toward foreign websites for online shopping.

In conclusion, the author argues that attitude toward foreign websites for buying products online might be influenced by perceived risks. However, this research cannot show any statistical evidence for these inferences because of the small sample size and the limitations of the questionnaire items used in this study. Scales to measure the concepts have to be well approved before conducting large-scale studies.

REFERENCES

[1] J. J. Kacen, J. D. Hess, and W.-y. K. Chiang, "Bricks or clicks? Consumer attitudes toward traditional stores and online stores", Global Economics and Management Review, vol. 18(1), pp. 12–21, 2013 .

[2] D. A. Karayanni, "Web-shoppers and non-shoppers: compatibility, relative advantage and demographics", European Business Review, vol. 15(3), pp. 141–152, 2003.

[3] G. L. Geissler and G. M. Zinkhan, "Consumer Perceptions of the World Wide Web: An Exploratory Study Using Focus Group Interviews", Advances in consumer research, 25(1), pp. 386-392, 1998.

[4] J. B. Schafer, J. A. Konstan, and J. Riedl, "E-commerce recommendation applications", in Applications of Data Mining to Electronic Commerce: Springer, pp. 115–153, 2001.

[5] P. Jiang and D. B. Jones, "An Exploratory Study of Factors Affecting Consumer International Online Shopping Behavior", International Journal of E-Business Research, vol. 10(2), pp. 45–60, 2014, doi:10.4018/ijebr.2014040103

[6] H. Park, L. Davis Burns, and N. J. Rabolt, "Fashion innovativeness, materialism, and attitude toward purchasing foreign fashion goods online across national borders" Journal of Fashion Marketing and Management: An International Journal, vol. 11(2), pp. 201–214. 2007, doi:10.1108/13612020710751383

[7] S. Ha and L. Stoel, "Consumer e-shopping acceptance: Antecedents in a technology acceptance model", Journal of Business Research, vol. 62(5), pp. 565–571, 2009, doi:10.1016/j.jbusres.2008.06.016

[8] C. Park and Y. Kim, "Identifying key factors affecting consumer purchase behavior in an online shopping context", International Journal of Retail & Distribution Management, vol. 31(1), pp. 16–29, 2003. doi:10.1108/09590550310457818

[9] L. R. Vijayasarathy, "Predicting consumer intentions to use on-line shopping: The case for an augmented technology acceptance model", Information & management, vol. 41(6), pp. 747–762. 2004, doi:10.1016/j.im.2003.08.011

[10] T. F. Stafford, A. Turan, and M. S. Raisinghani, "International and Cross-Cultural Influences on Online Shopping Behavior", Journal of Global Information Technology Management, vol. 7(2), pp. 70–87, 2004.

[11] T. P. Liang and J. S. Huang, "An empirical study on consumer acceptance of products in electronic markets: a transaction cost model", Decision support systems, vol. 24(1), pp. 29–43, 1998.

[12] ATKearney. *Global Retail E-Commerce Keeps On Clicking: The 2015 Global Retail E-Commerce Index.* [Online] Available from: https://www.atkearney.com/ 2017.08.30

[13] eMarketer. *Retail e-commerce sales worldwide from 2014 to 2021.* [Online]. Available from: https://www.emarketer.com 2017.08.30

[14] PWC. *Total Retail Survey 2016.* [Online]. Available from: https://www.pwc.com 2017.08.30

[15] M. Limayem, M. Khalifa, and A. Frini, "What makes consumers buy from internet? A longitudinal study of online shopping", IEEE Transactions on Systems, Man, and Cybernetics, vol. 30(4), pp. 421–432, 2000.

[16] H. van der Heijden, T. Verhagen, and M. Creemers, "Understanding online purchase intentions: contributions from technology and trust perspectives", European journal of information systems, vol. 12(1), pp. 41–48, 2003.

[17] M. K. Chang, W. Cheung, and V. S. Lai, "Literature derived reference models for the adoption of online shopping", Information & management, vol. 42(4), pp. 543–559. 2005, doi:10.1016/j.im.2004.02.006

[18] S. L. Jarvenpaa and P. A. Todd, "Is there a future for retailing on the Internet" Electronic marketing and the consumer, vol. 1(12), pp. 139–154, 1997.

[19] N. Li and P. Zhang, "Consumer online shopping attitudes and behavior: An assessment of research", AMCIS 2002, vol. 74, pp. 508-517, 2002.

[20] L. Zhou, L. Dai, and D. Zhang, "Online Shopping Acceptance Model - A Critical Survey of Consumer Factors in Online Shopping", Journal of Electronic Commerce Research, vol. 8(1), pp. 41–62, 2007.

[21] D. Cyr, "Modeling Web Site Design Across Cultures: Relationships to Trust, Satisfaction, and E-Loyalty", Journal of Management Information Systems, vol. 24(4), pp. 47–72, 2008, doi:10.2753/MIS0742-1222240402

[22] B. Ganguly, S. B. Dash, and D. Cyr, "The effects of website design on purchase intention in online shopping: the mediating role of trust and the moderating role of culture", International Journal of Electronic Business, vol. 8(4/5), pp. 302–330, 2010.

[23] W. Gong, L. M. Maddox, and R. L. Stump, R. L, "Attitudes toward online shopping: A comparison of online consumers in China and the US", International Journal of e-business development, pp. 28-35, 2012.

[24] H. Cho, M. Rivera-Sanchez, M., and S. S. Lim, "A multinational study on online privacy: Global concerns and local responses", New Media & Society, vol. 11(3), pp. 395–416, 2009. doi:10.1177/1461444808101618

[25] C. Yoon, "The effects of national culture values on consumer acceptance of e-commerce: Online shoppers in China", Information & management, vol. 46(5), pp. 294–301, 2009. doi:10.1016/j.im.2009.06.001

[26] D. Gefen and T. Heart, "On the need to include national culture as a central issue in e-commerce trust beliefs", Journal of Global Information Management, vol. 14(4), p. 1, 2006.

[27] J. Nantel and E. Glaser, "The impact of language and culture on perceived website usability", Journal of Engineering and Technology Management, vol. 25(1-2), pp. 112–122, 2008. doi:10.1016/j.jengtecman.2008.01.005

[28] X. Tong, "A cross-national investigation of an extended technology acceptance model in the online shopping context", International Journal of Retail & Distribution Management, vol. 38(10), pp. 742–759, 2010. doi:10.1108/09590551011076524

[29] N. Singh, G. Fassott, M. C. Chao, and J. A. Hoffmann, "Understanding international web site usage", International Marketing Review, vol. 23(1), pp. 83–97, 2006. doi:10.1108/02651330610646304

[30] W. Hwang, H.-S. Jung, and G. Salvendy, "Internationalisation of e-commerce: A comparison of online shopping preferences among Korean, Turkish and US populations", *Behaviour & Information Technology*, vol. *25*(1), pp. 3–18, 2006. doi:10.1080/01449290512331335636

[31] P. Norden, *A report on Norwegians habits and attitudes towards distance trading*. [Online]. Available from: http://hugin.info/ 2017.08.30

[32] K. B. Murray, "A test of services marketing theory: consumer information acquisition activities", *The journal of marketing*, pp. 10–25, 1991.

[33] H. Ko, J. Jung, J., J. Kim, and S. W. Shim, "Cross-cultural differences in perceived risk of online shopping", Journal of Interactive Advertising, vol. 4(2), pp. 20–29, 2004.

[34] A. D. Miyazaki and A. Fernandez, "Consumer Perceptions of Privacy and Security Risks for Online Shopping", Journal of Consumer Affairs, vol. 35(1), pp. 27–44, 2001. doi:10.1111/j.1745-6606.2001.tb00101.x

[35] S. M. Forsythe and B. Shi, "Consumer patronage and risk perceptions in Internet shopping", Journal of Business Research, vol. 56(11), pp. 867–875, 2003.

[36] P. A. Pavlou, "Consumer acceptance of electronic commerce: Integrating trust and risk with the technology acceptance model", International journal of electronic commerce, vol. 7(3), pp. 101–134, 2003.

[37] J. Cho, "Likelihood to abort an online transaction: influences from cognitive evaluations, attitudes, and behavioral variables", Information & management, vol. 41(7), pp. 827–838, 2004. doi:10.1016/j.im.2003.08.013

[38] T. S. H. Teo and J. Liu, "Consumer trust in e-commerce in the United States, Singapore and China". Omega, vol. 35(1), pp. 22–38, 2007.

[39] M. H. M. Javadi, H. R. Dolatabadi, M. Nourbakhsh, A. Poursaeedi, and A. R. Asadollahi, "An analysis of factors affecting on online shopping behavior of consumers", International Journal of Marketing Studies, vol. 4(5), p. 81, 2012.

[40] A. Borchers, "Trust in Internet shopping: A test of a measurement instrument", AMCIS 2001, p. 156, 2001.

[41] E. M. Rogers, Diffusion of innovations. Simon and Schuster, 2010.

[42] C. Park and J.-K. Jun, "A cross-cultural comparison of Internet buying behavior: Effects of Internet usage, perceived risks, and innovativeness", *International Marketing Review*, vol. *20*(5), pp. 534–553, 2003.

[43] J.-W. Lian and T.-M. Lin, "Effects of consumer characteristics on their acceptance of online shopping: Comparisons among different product types", *Computers in Human Behavior*, vol. *24*(1), pp. 48–65, 2008. doi:10.1016/j.chb.2007.01.002

[44] A. Varma Citrin, D. E. Sprott, S. N. Silverman, and D. E. Stem, "Adoption of Internet shopping: The role of consumer innovativeness", *Industrial Management & Data Systems*, vol. *100*(7), pp. 294–300, 2000. doi:10.1108/02635570010304806

[45] R. E. Goldsmith, "Using the Domain Specific Innovativeness Scale to identify innovative Internet consumers", *Internet Research*, vol. *11*(2), pp. 149–158, 2001. doi:10.1108/10662240110695098

[46] I. Ajzen, "The theory of planned behavior", *Organizational behavior and human decision processes*, vol. *50*(2), pp. 179–211, 1991.

[47] V. Venkatesh and F. D. Davis, "A theoretical extension of the technology acceptance model: Four longitudinal field studies", *Management science*, vol. *46*(2), pp. 186–204, 2000.

[48] S. Forsythe, C. Liu, D. Shannon, and L. C. Gardner, "Development of a scale to measure the perceived benefits and risks of online shopping", *Journal of interactive marketing*, vol. *20*(2), pp. 55–75, 2006.

[49] W. R. Swinyard and S. M. Smith, "Why people (don't) shop online: A lifestyle study of the internet consumer", *Psychology & marketing*, vol. *20*(7), pp. 567–597, 2003.

[50] R. Agarwal and J. Prasad, "A conceptual and operational definition of personal innovativeness in the domain of information technology", *Information systems research*, vol. 9(2), pp. 204–215, 1998.

[51] C. Fornell and D. F. Larcker, "Evaluating structural equation models with unobservable variables and measurement error", *Journal of marketing research*, pp. 39–50, 1981.

[52] J. F. Hair, M. Sarstedt, C. M. Ringle, and J. A. Mena, "An assessment of the use of partial least squares structural equation modeling in marketing research", *Journal of the academy of marketing science*, vol. *40*(3), pp. 414–433, 2012.

[53] J. Moon, D. Chadee, and S. Tikoo, "Culture, product type, and price influences on consumer purchase intention to buy personalized products online", *Journal of Business Research*, vol. *61*(1), pp. 31–39, 2008. doi:10.1016/j.jbusres.2006.05.012.

TABLE V. CORRELATION AMONG VARIABLES

Variables	*2*	*3*	*4*	*5*	*6*	*7*	*8*	*9*	*10*	*11*	12	13
1 SXP	0.220	0.330**	0.335**	0.243	-0.058	0.009	-0.322*	0.152	-0.444**	0.364**	0.199	0.091
2 SUS	1	-0.172	0.190	0.118	-0.096	0.097	-0.240	-0.030	-0.187	0.367**	0.439**	0.203
3 AGE		1	0.250	0.330*	0.008	-0.019	0.003	-0.085	-0.138	-0.098	-0.036	-0.306*
4 GEN			1	0.207	0.030	-0.058	-0.016	0.141	-0.203	0.093	-0.031	0.126
5 EDU				1	0.125	-0.075	0.042	0.085	-0.247	0.000	-0.102	0.028
6 FIN					1	0.299*	0.451**	0.262*	0.143	-0.265*	-0.099	0.094
7 PRO						1	0.224	0.310*	-0.071	0.069	-0.068	-0.014
8 CON							1	0.206	0.133	-0.521**	-0.307*	-0.211
9 DEL								1	-0.152	-0.049	-0.280*	0.305*
10 RET									1	-0.126	0.132	-0.141
11 PIN										1	0.236	0.164
12 ADV											1	0.067
13 ATF												1

Notes: ** p<.01; * p<.05;
SXP=online shopping experience in years, SUS=online shopping frequency, AGE=age, GEN=gender dummy variable (male was coded 1), EDU=education, FIN=financial risk, PRO=product risk, CON=convenience risk, DEL=delivery risk, RET= return policy risk, PIN=personal innovativeness, ADV=relative advantage, ATF=attitude toward foreign websites for buying products online

Influence of the Perceived Data Security and Trust on Usage Frequency of Internet Services

Erik Massarczyk, Peter Winzer
Faculty of Design – Computer Science – Media
RheinMain University of Applied Sciences
Wiesbaden, Germany
Email: erik.massarczyk@hs-rm.de, peter.winzer@hs-rm.de

***Abstract*—An increasing customer usage of Internet services with various devices demands a greater effort on data security credibility and trust issues because the extensive connections personal data are spread more widely. However, customers often prefer better services rather than higher data security. Here, the aim of this paper is to examine the positive influence of the perceived data security on the usage frequency of Internet services. The main target will be to measure how the user perceived data security and perceived trust influences the usage frequency of Internet services. This will be analyzed with an adjusted conceptual model based on elements of the Unified Theory of Acceptance and Use of Technology 2. Generally, a significant positive influence of a perceived data security on the usage frequency for specific services can be found. Yet, the perceived trust in the service providers does not significantly relate to a stronger usage frequency of Internet services. Consequently, customers have data security concerns and these might hinder them to use several Internet services.**

Keywords-data security; trust; usage frequency; Internet services.

I. INTRODUCTION

The growth of the number of Internet services and of the number of users lead to an increased amount of gained data. Especially services like (a) instant messaging, (b) social media, (c) video on demand (broadcasting/streaming), (d) gaming, and (e) cloud computing are used by more and more people with more different devices [1][2][3]. Due to this application of services, the degree of connection of the people and devices increases quite heavily [1]. Based on the growth of the number of connections and Internet services usages, the users produce more personal data and the data is spread to a larger degree [2].

From the customer point of view, it is difficult to comprehend to which extent personal data is collected, where the personal data is stored and which persons get access to handle the raised personal data for legal or illegal motives [3][4]. Due to the increased connectivity between the devices, unhindered individual communications and marketing measures, a wide range of information and personal data is disclosed. The data disclosure touches the security and privacy concerns of the customers because the personal information could include critical information and intellectual properties of the users themselves. Furthermore, personal information are countable assets from which enterprises and criminals can benefit [1][5]. Nonetheless, each user is responsible which data he or she releases for the usage of the specific Internet services and the different devices. Obviously, a lot of people are willing to distribute their personal information to get a good performance of the used services. Here, they often do not care about risks of data leakages and data misuse.

The rising number of security incidents shows that criminals more frequently attack enterprises, administrations and private customers to get the personal data because they have detected the values of these personal information and intellectual properties [6]. As a result, customers should care more about possible data privacy and security concerns, while using Internet services.

As a consequence, we want to examine if the private customers have data privacy and data security concerns when they use different Internet services with various devices. Together with the different conditions of wired and wireless networks and connections, different types of data security problems could arise. In this respect, we want to measure the status and the perception of data security while customers using the following services: (a) email, (b) social media, (c) online telephony, (d) online shopping, (e) cloud computing, (f) e-learning, (g) instant messaging, (h) online banking, (i) navigation, (j) online administration, (j) video on demand, and (k) internet television. Additionally, we add the customer evaluation of the trust of the providers of the named Internet services. Here, it will be measured how the customers perceive that the providers of the Internet services in general further distribute their personal data. Due to customers use the named Internet service differently in the wired and wireless networks, we separate the results in the two named considerations. On the hand, we consider the perceptions in the fixed/wired infrastructure environment and on the other hand, the results in the mobile/wireless infrastructure environment. Therefore, we implement a variable how the customers perceive the credibility of the network security (operator).

This implementation should just show how the customers estimate the network, but it would not be used for the analysis of the relation to the usage frequency of Internet services.

For this reason, the perception of the credibility of the operator and the importance of data security are falling behind the major considerations of the customer perception of data security by using Internet services and the trust provider of these services.

In Section II, the term data security, the known literature and used research models will be described. Following this section, the methodology, as well as the theoretical approach for carrying out the analysis, will be briefly explained. In Section IV the results of the hypothesis tests are briefly presented. Finally, in Section V, a critical discussion of the results takes place.

II. Literature Review

A. Data Security

In general, the term "data security" describes the secure management of personal data, secure data transmission and the transparency of which institutions or persons have access to the personal customer data [5][7]. The correct implementation of data security usually involves that the customers themselves decide who is entitled to access their data. As mentioned in the introduction, customers often ignore possible risks of sharing information and they are not aware of the amount of data, which they produce and which are the consequences if the personal data would be leaked [8][9][10]. The ignorance shows critical issues in three dimensions. Firstly, customers spread personal data which could be linked to confidential information like bank accounts and credit card numbers [8][9]. Secondly, a lot of companies use and transmit – without permission and knowledge of the customers – private customer information, which the customers disclose during the usage of Internet services [11]. Thirdly, as already mentioned, the number of Internet security incidents – like criminal acts of password capturing, eavesdropping and blackmails – have increased quite heavily during the last couple of years [3][6].

Yet, the perceptions of (a) data security, (b) trust, (c) credibility, (d) sharing of information and (e) risks differs between the individual customers and depend beside others on factors like demography and culture [8]. It is also known that most of the customers prefer a good Internet service performance instead of strong security or data protection measures. Here, customers frequently do not care about the consequences of misuse and data leakage. Especially these behaviors motivate us to investigate which factors directly influence the usage frequency of Internet services and the individual perception of data security and trust.

B. Research Model – Adjusted Model with Elements of the Unified Theory of Acceptance and Use of Technology 2

The main target of this study will be to get an increased comprehension of private customer behaviors, especially in the focus on data security and trust concerns and the acceptance and actual usage of services.

The Unified Theory of Acceptance and Use of Technology 2 (UTAUT2) is the direct expansion of the known UTAUT concepts with the factors hedonic motivation, price, and habit/experience, which allows a broader consideration of critical influence factors on user behavior [12]-[15].

Nevertheless, perceived data security and perceived trust could not be covered by the existing variables of UTAUT2. Nonetheless, an implementation of external variables as influence factors of the user behavior could be performed. By the approach of Escorbar-Rodriguez and Carvajal-Trujillo, the UTAUT2 model could be expanded by external variables trust as well as the further components perceived security and perceived privacy [12][16]. This expansion makes clear that the influence of security measures and perceptions on the behavioral intention to use of an innovation can be investigated [12][16]. Furthermore, this approach motivates us to use the factors perceived data security and perceived trust as external variables in the own adapted model (see Figure 1) [16]. Therefore, the adapted model keeps only the basic idea of the UTAUT2. In this context, Lin et al. have figured out that data security and privacy are the most affecting factors for an acceptance and adoption of a new technology [13].

Consequently, we want to directly measure the impact of the perceived data security measures on the actual usage of Internet services (instead of testing the relationship with the behavioral intention to use, as Zhong et al. already did [17]). In other words: The target of investigation is to analyze whether perceived data security and trust issues lead to a utilization of an Internet service.

Generally, we estimate that an increased perception for data security measures and trust concerns would lead to an increased usage of services. For the further combined regression analyses, the external variables perceived credibility (operator credibility) and importance of data security are also implemented. Perceived credibility describes the users' belief that the used systems would be free of threats for privacy and security and how the customers estimate and perceive the reliability of the service providers [15]. Customers recognize the behavior of providers if they take care (or if they don´t do so) about the personal information and secure transmissions [17]-[27].

Due to the fact that using Internet services (especially mobile services) include security and privacy threats [18], we implement the factor trust. The perception of trust describes how credible the customers perceive the provider [1][16][28][29][30]. Based on the assumption that risks and perceived trust directly influence the usage processes [31], the customers would reduce their usage if they expect a loss of privacy and a higher risk in usage [1][32][33][34]. The particular importance of the key factors of risk and trust lies in the fact that these two factors have a major influence on the customer acceptance of innovations (especially mobile payments, mobile banking and mobile shopping) [17][18][35]-[38]. In addition, trust in a service or in a service provider plays an important role for the customer, since this increases the customer's sense of satisfaction in the service and thus leads to a higher usage frequency [31][39].

Finally, non-existent trust or the perception of missing security negatively impact customer behavior. An increase in security by using a service would give the customers a more confident, secured and satisfied emotion and could possibly imply a stronger usage of this service. For this reason, the used survey also includes questions about how the customers perceive the security of the infrastructure and how the network operators use the gained data from the customer.

Based on these explanations, the hypotheses for this research paper are:

H1: The customer perception of data security has a directly positive effect on the usage of Internet services.

H2: An increased perceived provider trust has a directly positive effect on the usage of Internet services.

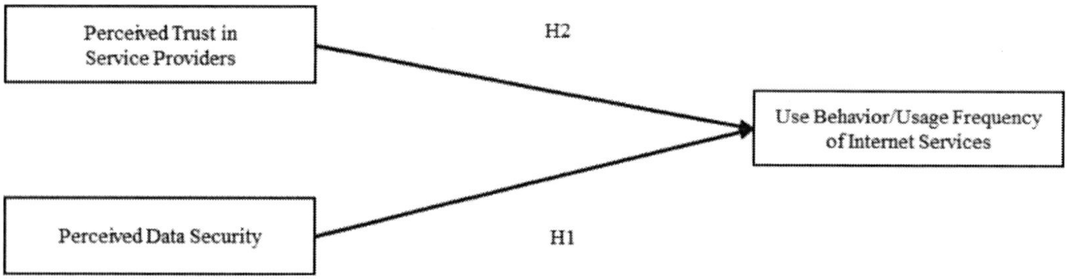

Figure 1. Conceptual Model

III. METHODOLOGY

The hypotheses are validated on the basis of a current survey. The answers were taken by interviewers in personal interviews, thus ensuring completeness and accuracy of the answers. The respondents were randomly chosen and asked if they wanted to answer the questionnaire. The interviewers were instructed to choose the interviewees as far as possible randomly to make sure to get a sample which represent the demographic characteristics of gender and age of the local population [40][41]. Generally, test persons are asked in December 2016 at public libraries in Wiesbaden (which is a city with approx. 275,000 inhabitants in the middle of Germany) to reach a diversified and representative selection of test persons. In total, the survey includes 290 completed questionnaires. The collected data has been examined based on quantitative research methods with the statistical program Statistical Package for the Social Sciences (SPSS). To evaluate the reliability and validity of the obtained data, Cronbach Alpha was determined and an Exploratory Factor Analysis was performed.

The perceived data security was queried with the question, how the customers perceive their personal data for each specific Internet service in the usage of a fixed and/or mobile Internet access (5-Point-Likert-scale: very secure to very unsecure). For the measurement of the usage frequency of (mobile) Internet services, a 5-Point-Likert-scale (very often to very few) has been used [42]. Finally, the trust is measured by the question of whether or what users perceive the Internet service providers to spread their personal data (unauthorized).

As mentioned above, the used approach only keeps elements of the UTAUT2. Therefore, we do not follow the analysis with a Structural Equation Modeling. Instead, we use the ordinary least square regressions to test the significance of each of the named hypotheses [12][13]. In the afterwards following combined approach under recognizing and controlling of further variables like importance of data security, perceived credibility and password changing behavior, we use a combined regression analysis.

IV. DATA ANALYSIS AND RESULTS

A. Result Conditions

The following discussion assumes far predominantly that the participants of the survey answer as private customers, even if it cannot be completely excluded that some of the respondents may also answer from their perspective of personal small enterprises.

We will describe the results of the reliability and validity tests of the overall used hypotheses briefly. After this testing, the regression results of hypotheses will be prioritized to figure out the relationships between (a) perceived data security as well as perceived trust in the service providers and (b) the usage of specific Internet services.

B. Descriptive Results

It could be achieved 290 completed questionnaires. However, the expanded second survey covers 7 sets of questions. 55.0% of the respondents are male and the average

age of a respondent is between 30 and 39 years. With 48.1%, the group of the 20 and 29-year-olds has the largest share of respondents. Thus, this age group (which is 12.2% of the total population in Germany) is overrepresented in the survey by a factor of four [43]. Based on a study of ARD/ZDF from 2015 the 20 to 29-year-old nearly 100% Internet users [44].

The over-representation in younger age groups naturally leads to an under-representation of the elder age groups. Consequently, the collected data are not representative.

26.5% of respondents feel confident about their data, but on the contrary, 32.9% of respondents feel more or less insecure about their data. Interestingly, the one third of respondents, who feel insecure in their data security, does not fit at all with the results of the password changing behavior of the customers, since more than 80% of the customers change their passwords much less frequently than once a year: For email accounts 84.3% and for social media accounts 89.2%. Normally it would be expected that more people change their passwords more regularly if they will a data insecurity. In this respect, it can be stated as the first conclusion that the perception of the data security does not affect the frequency of the password changes. This could be the reason because a higher password security increases the overall security, but it does not affect the data privacy if the customers distribute their data on their own.

89.0% of respondents use an anti-virus program which fit with the quotas of 85.5%, which are also confirmed by studies by the software company McAfee, which reported 85.5% [45].

In average, the customers believe that fixed Internet providers have a little bit safer infrastructure than mobile Internet providers. Email services are the mostly used services overall (round about 80%). In the fixed infrastructures, about 3/4 of the customers use online shopping, video on demand and online banking (independent from the usage frequency). In the consideration of mobile devices and mobile infrastructures, about 4/5 of the customers use instant messaging.

TABLE I. IMPORTANCE OF DATA SECURITY.

Internet Services	Importance of Data Security
Email	54.9% very high importance
Social Media	31.9% very high importance
Online Shopping	53.6% very high importance
Online Banking	75.9% very high importance
Instant Messaging	47.4% very high importance

TABLE II. USAGE FREQUENCY.

Internet Services	Usage Frequency
Email	35.1% very frequently
Social Media	43.8% very frequently
Online Shopping	4.9% very frequently
Online Banking	6.9% very frequently
Instant Messaging	63.0% very frequently

The tables show the different services: (I) the importance of data security, (II) the usage frequency of Internet services, and (III) confidence in service providers. Interestingly, customers in the services they use very frequently (social media and instant messaging) feel a relatively low data security. Customers also recognize that the providers of these services do not particularly secure the customer data and use it for their own purposes. In opposite, the usage of online banking is relatively rare, but data security is very important to customers in this area, which is, of course, mainly due to the nature of the service and is presumably independent of the channel through which this financial service is provided.

C. *Reliability and Validity*

The results of the reliability and validity analyses are illustrated in the Tables IV and V. In general, this study includes the following 7 aspects: (1) usage of Internet services (fixed networks), (2) usage of Internet services (mobile networks), (3) usage frequency of Internet services, (4) perceived importance of data security, (5) perceived data security (fixed networks), (6) perceived data security (mobile networks), and (7) perceived trust.

Generally, all named concepts are examined in the terms of reliability and validity. Following Cronbach, Alpha values must be higher than 0.7 to for a good reliability [46][47][48]. Based on the results in Table IV, the collected data for the 7 named aspects are reliable.

After the testing of the reliability, the exploratory factor analysis includes the assessment of Kaiser-Meyer-Olkin criterion (KMO), the significance test from Bartlett, and the examination of the cumulative variance to evaluate the validity of the collected data [49]-[53]. To reach a good validity, the concepts should reach significant p values ($p<0.05$) in the Bartlett-Test and KMO values above 0.7 [49]-[53].

Table V shows good validity scores of the collected data/aspects can be comprehended. The good validity scores are also supported by the results of the cumulative variances higher than 50%, which indicate high explanation rates of the collected data [50][51][52]. Consequently, the reliability and validity of the collected data are proved.

TABLE III. TRUST IN SERVICE PROVIDERS

Internet Services	Trust in Data Usage – closed	Trust in Data Usage – open
Email	15.7% very closed	3.9% very open
Social Media	2.1% very closed	21.1% very open
Online Shopping	5.1% very closed	14.1% very open
Online Banking	45.0% very closed	1.7% very open
Instant Messaging	4.6% very closed	14.9% very open

TABLE IV. RELIABILITY ANALYSIS

Research Concepts	Cronbach's Alpha
Usage of Internet Services (fixed networks)	0.780
Usage of Internet Services (mobile networks)	0.784
Usage Frequency of Internet Services	0.803
Perceived Importance of Data Security	0.925
Perceived Data Security (in fixed infrastructures)	0.881
Perceived Data Security (in mobile infrastructures)	0.915
Perceived Trust	0.871

TABLE V. VALIDITY ANALYSIS

Research Concepts	KMO	Bartlett -Test	Cumulative Variance
Usage of Internet Services (fixed)	0.825		50.397%
Usage of Internet Services (mobile)	0.804		51.240%
Usage Frequency of Internet Services	0.781	p	53.724%
Perceived Importance of Data Security	0.901	< 0.000	64.709%
Perceived Data Security (fixed)	0.844		57.791%
Perceived Data Security (mobile)	0.831		62.055%
Perceived Trust	0.827		59.372%

D. Regression Analyses

As mentioned above, the scope of the study does not allow the testing of all hypotheses.

In the following, at least, the relationship between the factors perceived data security, perceived trust and the usage frequency of Internet services will be analyzed by means of ordinary least square regressions. The perceived data security is analyzed differently for the use of fixed and mobile Internet services. This differentiation takes account of the fact that the various network / service types have different advantages and disadvantages, and therefore also different uses can be expected.

Following the named regression analyses, we combined all possible influence factors of security issues which they have collected in the survey to analyze their impact on the usage of Internet services.

For this purpose, the perceived data security (= independent variable) is analyzed separately for mobile and fixed broadband infrastructures / services) in relation to the usage frequency of the individual Internet services (= independent variables); see Table VI.

The r-square values of the individual regressions are quite low, which is mainly due to two causes. On the one hand, only the effects of perceived data security are analyzed for the usage frequency of each service. In each individual case, an r-square for the regression between only an independent variable and a dependent variable is determined. In so far as it is assumed, the individual r-squares are not quite as high. On the other hand, the usage frequency of an Internet service does not depend solely on the perceived data security. Based on the estimation of many different influencing factors (some are mentioned in the presented research model), the r-squares cannot be quite so high and we assume weak regressions.

For the usage of the following services in the fixed and mobile infrastructures, (a) Internet protocol television (IPTV), (b) instant messaging, and (c) online gaming, the customer data security perception does not impact the usage of these services; therefore, the hypothesis H1 cannot be accepted. For the services e-learning and cloud computing, significant positive regression relations could be found for both infrastructures (fixed and mobile). This means if a customer perceives a higher data security in his learning application, he will use the service more frequently. The coefficients of 0.286 (fixed) and 0.370 (mobile) show a quite moderate explanatory rate. As mentioned above, the r-squares of 3.3% (fixed) and 5.7% (mobile) are quite low and describe only a low coefficient of determination. Also, if customers perceive a higher data security when they use cloud services then they will use them more frequently. Coefficients of 0.330 (fixed) and 0.232 (mobile) and r-squares of 5.8% (fixed) and 2.8% (mobile) shows a moderate explanatory rate and low degree of determination [52][53][54]. For these both services, we do not assume differences in the usage of the services in the both infrastructures and the hypothesis H1 could be accepted.

The analysis of other services (online shopping, online banking, e-mail, social media, online telephony) shows differences in the results of the regression analyses between mobile or fixed infrastructures. The main reason for differences is the general use of services. Navigation and social media services are used by mobile devices in mobile infrastructures almost twice as frequently as fixed-line connections. In contrast, online banking services are used much more frequently via fixed broadband infrastructures than mobile connections.

The perceived data security has only a relatively small (but measurable) influence on the use of navigation services with mobile devices / networks only weak: regression of 0.161 and r-square of 2.1%. This may be due to the fact that the primary goal of most users of a navigation service is to locate a destination and it is self-evident to them that they may have to make concessions for data security (for example, by authorizing the location).

For fixed networks, positively significant regressions between the perceived data security for emails respectively perceived online banking data security and the usage of these services could be identified. Despite low r-squares of 5.8% (email) and 5.5% (online banking) and weak regressions, the single coefficients of 0.357 (email) and 0.295 (online banking) represent moderate explanatory rates [52][53][54].

Since e-mails and, in particular, bank accounts generally contain highly sensitive data from customers, the loss of which can cause considerable damage, customers' need for high data security for these services is, of course, particularly high. If the users perceive a better data security for these services, or if the service providers can guarantee their customers a higher data security, they will use these services more frequently.

TABLE VI. REGRESSION ANALYSIS – COMPARISON PERCEIVED DATA SECURITY AS INFLUENCE FACTOR FOR USAGE FREQUENCY (single service consideration)

Dependent variables	Independent: Perceived Data Security in Fixed Networks			Independent: Perceived Data Security in Mobile Networks		
	Regression Coefficient B	**Significance**	**R-Square**	**Regression Coefficient B**	**Significance**	**R-Square**
Usage Frequency of Email Services	0.357**	p<0.05	5.8%	No Significance		
Usage Frequency of Cloud Computing Services	0.330**	p<0.05	5.8%	0.232**	p<0.05	2.8%
Usage Frequency of Online Banking Services	0.295**	p<0.05	5.5%	No Significance		
Usage Frequency of E-Learning Services	0.286**	p<0.05	3.3%	0.370**	p<0.05	5.7%
Usage Frequency of Instant Messaging Services	No Significance			No Significance		
Usage Frequency of IPTV Services	No Significance			No Significance		
Usage Frequency of Navigation Services	No Significance			0.161**	p<0.05	2.1%
Usage Frequency of Social Media Services	No Significance			No Significance		
Usage Frequency of Online Gaming Services	No Significance			No Significance		
Usage Frequency of Online Administration Services	0.393**	p<0.05	6.7%	No Significance		
Usage Frequency of Online Shopping Services	No Significance			0.142*	p<0.05	1.8%
Usage Frequency of Online Telephony Services	0.228**	p<0.05	2.1%	No Significance		

* The regression presents a significant constant, which could be an indicator for further unconsidered variables or an existing endogeneity, which needs further investigation.

** The regression presents a significant constant, which could be an indicator for further unconsidered variables or an existing endogeneity, which needs further investigation. Furthermore, the Durban-Watson-Test recognizes a value which could be an indicator for an existing autocorrelation. To cover the spurious correlations, further investigations must be performed.

In addition, e-mail services are often used in professional contact and can contain corresponding confidential information [8] [9].

In general, the test of multicollinearities with the Variance Inflation Factor (VIF) shows that all VIF values are below 10 (mostly below 3) and therefore, multicollinearities not exist [49][55][56]. Nonetheless, in some cases, the constants are also significant ($p<0.05$), which could be an indicator for other influence factors or an existing endogeneity. In the further research and examination of the data, we will consider the influence factors and try to figure out which are the indicators for the significant constants.

The relationship of the perceived trust in the service providers (independent variable) and the usage frequency of Internet services (dependent variable) generally show no significant relationship for the specific services. The only exception is the service online shopping. The positive significant relationship (coefficient = 0.117) shows that customers, who perceive that the shopping providers do not further distribute their personal information, will more frequently use these online shopping platforms. However, the r-square of 1.7% and the coefficient below 0.200 do not imply a good explanatory rate and it must be assumed that regressive connection is weak [52][53][54]. Generally, the hypothesis H2 about the influence of the customer perception of trust in the service providers of the single specific Internet services on the usage frequency of the specific services cannot be accepted. It must be assumed that trust as single factor does not have an influence on the customer decision of service usage.

Finally, as already mentioned, a combined regression analysis approach is carried out with the implementation of all of the above concepts in order to analyze the influence on the frequency of the user behavior of the specific Internet services. The following variables are controlled: (a) overall perceived data security (in general without any consideration of a single service), (b) perceived importance of data security, (c) perceived credibility of the network operators and (d) perceived trust in the service providers. The regression analyses for each individual service are carried out separately and shown according to the use of mobile or fixed network services.

The control of the variables which cover security issues (except perceived data security) reveals significant regressive influences of perceived data security on the usage frequency of the specific Internet services (email, cloud computing, online banking and e-learning); see Table VII.

TABLE VII. REGRESSION ANALYSIS – COMPARISON OF DATA SECURITY AS INFLUENCE FACTOR FOR USAGE FREQUENCY
(combined independent variables consideration on single service consideration)

Dependent variables	Independent: Perceived Data Security in Fixed Networks*			Independent: Perceived Data Security in Mobile Networks*		
	Regression Coefficient B	Significance	R-Square	Regression Coefficient B	Significance	R-Square
Usage Frequency of Email Services	0.363***	$p<0.05$	9.7%	No Significance		
Usage Frequency of Cloud Computing Services	0.261	$p<0.05$	8.2%	No Significance		
Usage Frequency of Online Banking Services	0.218	$p<0.05$	12.0%	0.352**	$p<0.05$	17.7%
Usage Frequency of E-Learning Services	No Significance			0.328	$p<0.05$	14.5%

* Other independent variables overall perceived data security, perceived importance of data security, perceived credibility of the network operators and perceived trust in the service providers are controlled and implemented.

** The regression presents a significant constant, which could be an indicator for further unconsidered variables or an existing endogeneity, which needs further investigation.

*** The regression presents a significant constant, which could be an indicator for further unconsidered variables or an existing endogeneity, which needs further investigation. Furthermore, the Durban-Watson-Test recognizes a value which could be an indicator for an existing autocorrelation. To cover the spurious correlations, further investigations must be performed.

The control of the variables confirms the results obtained in the first point. When customers use e-mail services over the fixed networks and they feel confident about their data, they will use the data more frequently. Although nearly 80% of the customers use email services over the mobile networks, no significant connection could be found. Despite the non-significance for mobile networks, the regression coefficient of 0.363 for fixed networks shows a moderate explanatory rate [52][53][54]. However, the r-square of 9.7% describes only weak regression with a low coefficient of determination [52][53][54]. The VIF is below 3, so multicollinearities can be excluded [49][55][56]. It can be assumed that customers who experience more data security when using e-mail services will use these services more frequently. This is mainly because customers have stored many confidential information in their e-mail accounts and do not want third parties to have access to these data.

A similar relationship exists for cloud computing: when customers perceive higher data security for cloud computing services, they will use these services more frequently (significantly positive). Despite a moderate regression coefficient of 0.261, the r-square of 8.2% shows a weak regression. The VIF under 3 allows the exclusion of multicollinearities [49][55][56].

The third line of Table VII represents the influence of the perceived data security on the usage of online banking. Independently if the customers use online banking in the mobile or fixed networks, it can be identified that users, who have security issues with online banking, do not use online banking. The regression coefficients of 0.218 (fixed) and 0.352 (mobile) describe also moderate explanatory rates.

The r-squares of 12.0% (fixed) and 17.7% (mobile) do not imply strong regressions, however, the values are two to three times higher than the r-squares, mentioned above (see Table VI). These both percentages describe how much the perceived data security declare the decision how often online banking will be used. When people use online banking services, they care about data security issues.

The service e-learning is not used by many customers. But, when customers use the service in the mobile environment, the decision to use is influenced by data security issues. The coefficient of 0.328 describes a moderate explanatory rate. The r-square of 14.5% is similar to the results of online banking. Despite a normally classified weak regression, this value is better than the results presented above.

To support the previous findings and to expand the results, we have executed a combined regression analysis with the controls of most of the used variables. For the named services, the hypothesis H1 can be accepted because data security concerns impact the decision to use a service (frequently). However, for the other services (social media, IPTV, online gaming, instant messaging), the hypothesis H1 has to be rejected because an impact of data security issues on the usage of these services could not be significantly proved.

V. Conclusions And Future Work

In this paper, we have analyzed indicators which influence the decision and usage frequency of Internet services. The focus of the publication is on the effects of perceived data security and perceived trust in the use decision and the usage frequency of Internet services.

In the first step, the influence of perceived data security or perceived trust on the usage frequency of certain internet services was examined. To support the results so far and to expand the results, we have conducted a combined regression analysis, focusing on the impact of the data security perceived by customers on the use of the services.

It could not be proved in general that security concerns and especially concerns in data security and trust in service providers lead to a reduced or an increased usage of the services. Nonetheless, some evidences and implications for specific services like email, online banking and e-learning exist. Customers who perceive that their data will be safe, use the service more frequently than customers, who feel uncertain. The main question is, why only some of the used services are influenced. We are in the opinion that these developments directly depend on the nature of the service. For example, Bank accounts and e-mails usually contain confidential information, the losses of which can have serious consequences for customers. In contrast, the use of services, such as IPTV merely reveals some information to individual preferences or behaviors. However, most people do not appreciate this information as so critical.

The second investigation focuses on the perceived trust in service providers. It examines how the transfer of customer data to third parties is evaluated. Interestingly, no evidences for the influence of the perceived trust on the usage of Internet services could be found. It must be predicted that data distributions by the service providers do not impact the user's decision to use a service. This non-existing relation could be explained by the fact that the most people focus on the performance and usability of the Internet services instead of the security, which is mentioned in the second section of this study. Furthermore, it must be assumed that the most people are not aware about these data distributions. Therefore, the rejection of this hypothesis is not surprisingly.

To get a better overview, the other relations between the security concerns of data security importance, perceived operator credibility and password changing behavior must be also considered. Consequently, further data analysis and research would be necessary to deepen the current findings.

REFERENCES

[1] E. Massarczyk and P. Winzer, "Influence of the Perception of Data Security and Security Importance on Customer Usage of Internet Services," International Journal On Advances in Internet Technology, Thinkmind Library (ISSN: 1942-2652), volume 10, numbers 1 and 2, 2017, pp. 1-22

[2] International Telecommunication Union (ITU), "ICT Facts & Figures – The world in 2015," May 2015, https://www.itu.int/en/ITU-D/Statistics/Documents/facts/ICTFactsFigures2015.pdf, [retrieved: 09.2017]

[3] P. W. Dowd and J. T. McHenry, "Network Security: It's Time to Take It Seriously," Computer (1998), vol. 31, issue 9, IEEE Xplore Digital Library, Sept. 1998, pp. 24-28.

[4] D. Desai, "Law and Technology – Beyond Location: Data Security in the 21st Century," Magazine Communications of the ACM (2013), vol. 56, issue 1, ACM, Jan. 2013, pp. 34-36.

[5] F. S. Ferraz and C. A. Guimarães Ferraz, "Smart City Security Issues: Depicting Information Security Issues in the Role of a urban environment," 7th International Conference on Utility and Cloud Computing, IEEE/ACM, 2014, pp. 842-846.

[6] Kaspersky Lab, "Damage Control: The Cost of Security Breaches," IT Security Risks Special Report Series, https://media.kaspersky.com/pdf/it-risks-survey-report-cost-of-security-breaches.pdf., 2015, [retrieved 09.2017]

[7] D. Nayak, N. Rajendran, D. B. Phatak, and V. P. Gulati, "Security Issues in Mobile Data Networks," Vehicular Technology Conference (VTC 2004), vol. 5, IEEE Xplore Digital Library, Sept. 2004, pp. 3229-3233.

[8] S. Dhawan, K. Singh, and S. Goel, "Impact of Privacy Attitude. Concern and Awareness on Use of Online Social Networking," 5th International Conference - Confluence The Next Generation Information Technology Summit 2013, IEEE Xplore Digital Library, Sept. 2014, pp. 14-17.

[9] D. Malandrino, V. Scarano, and R. Spinelli, "How Increased Awareness Can Impact Attitudes and Behaviors Toward Online Privacy Protection," International Conference on Social Computing, IEEE Xplore Digital Library, Sept. 2013, pp. 57-62.

[10] A. N. Joinson. U.-D. Reips, T. Buchanan, and C. B. Paine Schofield, "Privacy, Trust, and Self-Disclosure Online, " Human-Computer Interaction, vol. 25, no. 1, 2010, pp. 1-24.

[11] B. Krishnamurthy, K. Naryshkin, and C. Wills, "Privacy Leakage vs. Protection Measures: the Growing Disconnect," in Web 2.0 Security and Privacy Workshop, 2011, pp. 1-10.

[12] V. Venkatesh, J. Y. L. Thong, and X. Xin, "Consumer Acceptance and Use of Information Technology: Extending the Unified Theory of Acceptance and Use of Technology," MIS Quarterly, vol. 36, issue 1, 2012, pp. 157-178.

[13] F.-T. Lin, H.-Y. Wu, and T. T. Nguyet Nga, "Adoption of Internet Banking: An Empirical Study in Vietnam," 10th International Conference on e-Business Engnineering, IEEE Xplore Digital Library, 2013, pp. 282-287.

[14] V. Venkatesh, J. Y. L. Thong, and X. Xin, "Consumer Acceptance and Use of Information Technology: Extending the Unified Theory of Acceptance and Use of Technology," MIS Quarterly, vol. 36, issue 1, 2012, pp. 157-178.

[15] V. Venkatesh, M. G. Morris, G. B. Davis, and F. D. Davis, "User Acceptance of Information Technology: Toward a Unified View," MIS Quarterly, vol. 27, issue 3, 2003, pp. 425-478.

[16] T. Escorbar-Rodriguez and E. Carvajal-Trujillo, "Online Purchasing Tickets for Low Cost Carriers: An Application of the Unified Theory of Acceptance and Use of Technology (UTAUT) Model," Tourism Management, vol. 43, 2014, pp. 70-88.

[17] J. Zhong, A. Dhir, M. Nieminen, M. Hämäläinen, and J. Laine, "Exploring Consumer Adoption of Mobile Payments in China," Academic Mind Trek'13, 2013, pp. 318-325.

[18] Y. S. Wang, Y. M. Wang, H. H. Lin, and T. I. Tang, "Determinants of User Acceptance of Internet Banking: an Empirical Study," International Journal of Service Industry Management, vol. 14, 2003, pp. 501-519.

[19] A. Zmijewska, E. Lawrence, R., and R. Steele, "Towards Understanding of Factors Influencing User Acceptance of Mobile Payment Systems," In: Proceedings of the IADIS WWW/Internet, Madrid, Spain, 2004.

[20] T. Dahlberg and A. Öörni, "Understanding Changes in Consumer Payment Habits – Do Mobile Payments and Electronic Invoices Attract Consumers?," In: 40th Annual Hawaii International Conference on System Sciences (HICSS), 2007, p. 50.

[21] P. G. Schierz, O. Schilke, and B. W. Wirtz, "Understanding Consumer Acceptance of Mobile Payment Services: An Empirical Analysis," Electronic Commerce Research and Applications, vol. 9, issue 3, 2010, pp. 209-216.

[22] C. Kim, W. Tao, N. Shin, and K. S. Kim, "An Empirical Study of Customers' Perceptions of Security and Trust in E-Payment Systems," Electronic Commerce Research and Applications, vol. 9, issue 1, 2010, pp. 84-95.

[23] K. Yang, "Exploring Factors Affecting the Adoption of Mobile Commerce in Singapore," Telematics and Informatics, vol. 22 issue 3, 2005, pp. 257-277.

[24] J. Cheong, M. Cheol, and J. Hwang, "Mobile Payment Adoption in Korea," In: ITS 15th biennial conference, Berlin, Germany, 2002.

[25] T. Dahlberg, N. Mallat, and A. Öörni, "Consumer Acceptance of Mobile Payment Solutions," In: G.M. Giaglis (ed.), mBusiness 2003 – The Second International Conference on Mobile Business, Vienna, 2003, pp. 211-218.

[26] N. Mallat, "Exploring Consumer Adoption of Mobile Payments – a Qualitative Study," Mobility Roundtable, Helsinki, Finland, vol. 16, issue 4, 2006, pp. 413-432.

[27] K. Pousttchi and M. Zenker, "Current Mobile Payment Procedures on the German Market from the view of Customer Requirements," In: 14th International Workshop on Database and Expert Systems Applications, 2003, pp. 870-874.

[28] R. De Sena Abrahao, S. N. Moriguchi, and D. F. Andrade, "Intention of Adoption of Mobile Payment: An Analysis in the Light of the Unified Theory of Acceptance and Use of Technology (UTAUT)," Innovation and Management Review, vol. 13, 2016, pp. 221-230.

[29] D. H. McKnight, L. L. Cummings, and N. L. Chervany, "Initial Trust Formation in New Organizational Relationships," Academy of Management Review, vol. 23, 1998, pp. 473-490.

[30] D. H. McKnight, V. Choudhury, and C. Kacmar, "The Impact of Initial Consumer Trust on Intentions to Transact with a Web Site: a Trust Building Model," The Journal of Strategic Information Systems , vol. 11, 2002, pp. 297-323.

[31] T. Zhou, "Understanding Mobile Internet Continuance Usage from the Perspectives of UTAUT and Flow," Information Development vol. 27, 2011, pp. 207-218.

[32] T. Zhou, Y. Lu, and B. Wang, "Integrating TTF and UTAUT to Explain Mobile Banking User Adoption," Computers in Human Behavior, vol. 26, 2010, 760-767.

[33] T. Zhou, "An Empirical Examination of Initial Trust in Mobile Banking," Information Development, vol. 21, issue 5. 2011, pp. 527-540.

[34] A. Y. L. Chong, "Understanding Mobile Commerce Continuance Intentions: An Empirical Analysis of Chinese Consumers," Journal of Computer Information Systems, 2013.

[35] L.-D. Chen, "A Model of Consumer Acceptance of Mobile Payment," International Journal of Mobile Communications, vol. 6, issue 1, 2008, pp. 32-52.

[36] M. A. Mahfuz, L. Khanam, and W. Hu, "The Influence of Culture on M-Banking Technology Adoption: An Integrative Approach of UTAUT2 and ITM," 2016 Proceedings of PICMET'16: Technology Management for Social Innovation, 2016, pp. 70-88.

[37] X. Luo, H. Li, J. Zhang, and J. P. Shim, "Examining Multi-dimensional Trust and Multi-faceted Risk in Initial Acceptance of Emerging Technologies: an Empirical Study of Mobile Banking Services," Decision Support Systems, vol. 49, issue 2, 2010, pp. 222-234.

[38] H.-P. Lu, and P. Y.-J. Su, "Factors Affecting Purchase Intention on Mobile Shopping Websites," Internet Research, vol. 19, issue 4, 2009, pp. 442-458.

[39] T. Oliveira, M. Faria, M. A. Thomas, and A. Popovic, "Extending the Understanding of Mobile Banking Adoption: When UTAUT meets TTF and ITM," International Journal of Information Management, vol. 34, 2014, pp. 689-703.

[40] J. Bortz and N. Döring, "Research Methods and Evaluations," [German] "Forschungsmethoden und Evaluation; für Human- und Sozialwissenschaftler," Heidelberg, Springer, vol. 4, 2009.

[41] M. Kaya, "Data Collection Procedure", [German] "Verfahren der Datenerhebung," in Albers, S./Klapper, D./Konradt, U./Walter, A./Wolf, J. (Hrsg.): Methodik der empirischen Forschung, Wiesbaden, Gabler, vol. 3, 2013, pp. 49-64.

[42] R. Likert, "A Technique for the Measurement of Attitudes," Archives of Psychology, 1932, pp. 199-224.

[43] Destatis, Statistisches Bundesamt, "Population," [German] "Bevölkerung," [Online] https://www.destatis.de/DE/ZahlenFakten/GesellschaftStaat/Bevoelkerung/Bevoelkerungsstand/Tabellen_/lrbev01.html, 2015, [retrieved 09.2017]

[44] Statista, "Internet Users in Germany from 2001 to 2015," [German] "Anteil der Internetnutzer in Deutschland in den Jahren 2001 bis 2015," [Online]http://de.statista.com/statistik/daten/studie/13070/umfrage/entwicklung-der-internetnutzung-in-deutschland-seit-2001/, 2015, [retrieved: 09.2017]

[45] Statista, "Customers without Anti-Virus Protection," [German], "Anteil der Verbraucher ohne aktives Antivirenprogramm in ausgewählten Ländern weltweit," https://de.statista.com/statistik/daten/studie/226942/umfrage/anteil-der-verbraucher-ohne-aktives-antivirenprogramm/, 2017, [retrieved: 09.2017]

[46] L. J. Cronbach, "Coefficient Alpha and the Internal Structure of Tests," Psychometrika, vol. 16, 1951, pp. 297-334.

[47] C. Fornell and D. Larcker, "Evaluating Structural Equation Models with Unobservable Variables and Measurement Error," Journal of Marketing Research, vol. 18, issue 1, 1981, pp. 39-50.

[48] R. Hossiep, "Cronbachs Alpha," [German] "Cronbachs Alpha," In Wirtz, M. A. (editor): Dorsch – Lexikon der Psychologie, vol. 17. Verlag Hans Huber, Bern, 2014.

[49] J. F. J. Hair, R. E. Anderson, R. L. Tatham, and W. C. Black, "Multivariate Data Analysis," Macmillan, New York, NY, Macmillan, vol. 3, 1995.

[50] S. Fromm, "Data Analysis with SPSS Part 1," [German] "Datenanalyse mit SPSS für Fortgeschrittene," Arbeitsbuch, vol. 2, VS Verlag für Sozialwissenschaften, GWV Fachverlage, Wiesbaden, 2008.

[51] S. Fromm, "Data Analysis with SPSS Part 2," [German] "Datenanalyse mit SPSS für Fortgeschrittene 2: Multivariate Verfahren für Querschnittsdaten," Lehrbuch, vol. 1, VS Verlag für Sozialwissenschaften, Springer, Wiesbaden, 2010.

[52] N. M. Schöneck and W. Voß, "Research Project," [German] "Das Forschungsprojekt – Planung, Durchführung und Auswertung einer quantitativen Studie," vol. 2. Springer Wiesbaden, 2013

[53] A. Field, "Discovering Statistics Using SPSS," Sage Publications Ltd., vol. 4, 2013.

[54] F. Brosius, "SPSS 8 Professional Statistics in Windows," [German] "SPSS 8 Professionelle Statistik unter Windows," Kapitel 21 Korrelation, International Thomson Publishing, vol. 1, 1998.

[55] D. Lin, D. P. Foster, and L. H. Ungar, "VIF Regression: A Fast Regression Algorithm for Large Data," Journal of the American Statistical Association, vol. 106, issue 493, 2009, pp. 232-247.

[56] S. Petter, D. W. Straub, and A. Rai, "Specifying Formative Constructs in Information Systems Research," MIS Quarterly, vol. 31, issue 4, 2007, pp. 623-656.

Quantifying Mobile User Experience

Status quo, Implementation Challenges, and Research Agenda

Maria Lusky and Stephan Böhm

CAEBUS Center of Advanced E-Business Studies
RheinMain University of Applied Sciences
Wiesbaden, Germany
Email: {maria.lusky, stephan.boehm}@hs-rm.de

Abstract—**In the last years, mobile applications (apps) have spread out to all parts of everyday life and the app market is growing rapidly. As developers and providers of mobile applications need to stay competitive in this environment, measuring the user experience (UX) of mobile applications is crucial in developing and maintaining mobile apps. In this context, methods for measuring user experience have to be applied to specific characteristics of mobile apps, such as context awareness, instable internet connections, small displays and alternative operating concepts. Against this background, this paper provides an overview of recent methods for measuring the user experience in the context of mobile applications and thus reveals research needs in this topic. We conduct a literature survey on UX related studies in the context of mobile applications of the last five years, taking into account generic methods that have been directly applied to the mobile context, methods that had been adapted, and new, mobile specific approaches. Furthermore, we propose a research agenda for the topic of mobile UX measuring.**

Keywords–User Experience (UX); Mobile User Experience; Human Computer Interaction (HCI); Mobile Applications; User Centered Design

I. INTRODUCTION

In the last years, mobile applications (apps) have been expanding to all parts of everyday life, including education, health, games, travel, shopping and work. The growing global app market is predicted to reach a total spend of over six trillion USD in 2021, which means an increase of almost four times its value [1]. According to a recent global study by the research agency App Annie, on a global average, one person uses nine apps per day and installs at least one new app per month [2]. Not only have mobile apps become an integral part in our everyday lives, but also the number of apps in the app stores is increasing. For every need, there are several alternative apps, and good user experience is needed to stay competitive. Moreover, smartphones as well as apps are becoming more complex, and challenge developers and publishers of mobile apps to sustain a good user experience. In this context, user experience does not only cover the usability of apps, but reaches out to emotional and motivational aspects of use. In order to generate a good user experience, user (experience) research needs to be a part of the conception, development and maintenance of a mobile app. Also, user analysis allows for insights in the users' habits and preferences. To reach this aim, there is a growing number of varying user experience evaluation methods for assessing different types of user experience data. Existing methods for user experience research are applied and adapted to the context of mobile applications, but there are also new approaches that have been developed specifically for this area.

Against this background, the main contribution of this work is on the one hand to provide an overview of recent methods for measuring the user experience in the specific context of mobile applications on smartphones and on the other hand to reveal research needs in this topic.

This paper is structured as follows: In Section II, the theoretical background of this work is outlined, covering general user experience models and models for measuring user experience. Section III focuses on the user experience in the context of mobile applications and smartphones. In Section IV, the results of the literature survey are presented. In Section V, the implications are discussed and challenges as well as research needs are pointed out based on the study findings. Section VI summarizes key findings and concludes this work.

II. THEORETICAL BACKGROUND

User experience (UX) in the field of human computer interaction describes end users' experiences on interacting with a system or service. The concept of user experience can be seen from different perspectives, as a phenomenon, a field of study, or a practise. According to Roto et al. [3], the field of study investigates the experiences a person can make and how they develop. The ways to design systems in order to create a particular UX and methods for assessing UX are also subject of this research. In [4], several definitions of UX from industry and academia have been collected. Based on this work, Roto et al. [3] have developed the following definition:

"The field of UX deals with studying, designing for and evaluating the experiences that people have through the use of (or encounter with) a system. This use takes place in a specific context, which has an impact on, or contributes to, the UX."

This definition underlies the work that is presented in this paper. In order to approach this topic, we will first give an overview of different user experience models and based on that introduce models for measuring user experience.

A. User Experience Models

There is a variety of user experience models, such as [5]–[8]. In this Section, we will give a brief overview of two models from user experience practise and the description of UX in an ISO standard. The standard EN ISO 9241 describes guidelines of human computer interaction. Following this standard, user experience, as described in EN ISO 9241-210, can be applied to three phases of interaction: previous to the interaction, during the interaction, and after the interaction. Furthermore, EN ISO 9241-11 describes usability as one aspect of user experience that is located in the phase during the interaction.

In [7], Garrett introduces his model *"the elements of user experience"*. According to this model, user experience comprises five levels *–strategy, scope, structure, skeleton, and surface–* that together form every digital product and, during the development process, have to be taken into account successively. At the first level of *strategy*, the purpose and goal of a product, as well as user requirements are specified. On the *scope* level, functional specifications and content requirements are determined. After that, the general *structure* is defined, comprising the information architecture and the interaction design that forms the background of the product's *skeleton*. On this next level, information, navigation and the interface are designed. At the top level, the *surface*, the sensory design of the product is made. While the first levels are rather abstact, the top levels are concrete and at all levels, the users' needs have to be taken into account.

Another, more comprehensive model is given by Stern [8]: The CUBI model conceptualizes user experience as four overlapping circles – *content, user goals, business goals,* and *interaction* (CUBI). Each of the circles is specified in five additional layers representig aspects that have to be considered for dealing with the particular field. The intersections of each two circles represent the four steps of a process that describes a user journey: *attraction, reactions, actions* and *transactions*. The intersections where three circles overlap are called *experience factors* and constitute primary factors of an effective user experience: *branded experience, comprehensive experience, useful experience* and *usable experience*. While the first model is more practically oriented and addressed to people who plan or develop products, the second one is theoretical and more suitable for business processes. However, both models have in common that UX is a concept that consists of several aspects where content, interaction, the users' needs and the goals of a product or service are closely intertwined. Each of these different aspects can be measured in order to assess different aspects of user experience.

B. Models for Measuring User Experience

There is a large variety of metrics and methods for measuring user experience, that are comprehensively described in user experience literature [9][10]. Prominent research in this field was done by Vermeeren et al. [11], who collected 96 methods from industry and academia and described how to use them. Though unified models for quantifying user experience are missing, there are several approaches to organize and classify existing evaluation methods. Vermeeren et al. [11] classify them regarding certain properties, such as the type of the collected data, the study type, or the development phase that the method can be used in.

Likewise, Albert and Tullis [9] established a categorization for different types of UX evaluation methods, distinguishing five types: (1) methods for assessing the performance, comprising task success, task completion time, occurring errors, efficiency, and learnability of the system; (2) issue-based methods, i.e., usability methods; (3) self-reported methods, such as rating scales and questionnaires; (4) behavioral and physiological methods, for example eye tracking, emotion tracking, heart rate and skin conductance; (5) combined methods. Our literature survey on UX evaluation methods aligned to the specifities of mobile apps and smartphones presented in the following sections will be oriented towards this classification.

Though Vermeeren et al. continued research in the field of UX measuring, as Law et al. [12], their work does not indicate, if the methods can be applied to the context of mobile applications on smartphones. However, this is a key issue for measuring UX, since mobile user experience differs from desktop user experience.

III. Mobile User Experience

To our understanding and in the scope of this work, mobile apps are application software to run on mobile devices, such as smartphones, with which the functionality given by hardware and operating software can be applied to solve user-specific problems. Typically, mobile apps consist of programs and data that will be installed by the end users themselves to the devices and thus are also an important element of handset personalization. One main characteristic of mobile applications compared to classic desktop software is that they are smaller and more specific. Different from mobile web usage, mobile apps can integrate a broader spectrum of smartphone hardware functionalities and interfaces, such as taking pictures with the camera, scanning a bar code or sending voice messages using the microphone. Another characteristic of mobile applications is that its sensoric functions can contextualize the usage situation with regard to current location, phone orientiation or other ambient conditions.

Hart [13] summarizes characteristics that distinguish mobile devices from desktop computers: Mobile phones have smaller screens with fewer pixels and therefore can display less information in a less detailed way. Also, smartphones are equipped with slower processors, making them less performant, and have access to less bandwidth than desktop computers. Rather than a mouse, they have a touch-based input and a small, hard to access keyboard, making them less precise and more challenging for text input. In addition, mobile phones often provide no or only limited multitasking, meaning that it is difficult to work with more than one app at once. Different from desktop browsers, with mobile phones websites can be run in browser applets inside an app, which leads to different functionalities and views while interacting with a web page. The portrait screens are another challenge for mobile application design, since they are unfavorable for displaying more than two colums or showing overly-wide elements. Thus, navigation in mobile apps is rather guided along the top than the side. Lastly, users are using mobile devices differently, in different settings, locations and situations that desktop computers, which has a crucial influence on the UX of mobile applications.

Due to these specifics, practitioners and academics in UX often differentiate between desktop UX and mobile UX, e.g., in [14]–[18]. There are several studies that are approaching the differences between desktop and mobile UX: Selke [19] points out that due to the smaller displays, reduced bandwidth and touch technology, users feel less comfortable while using smartphones or tablets than they do with desktop computers. Furthermore, mobile usage leads to different user behavior, such as a different search behavior, a lower rate of exploration in browsing, finishing different process steps in a task, and receiving, reading and understanding a different amount of information [19]. Additionally, the UX with using mobile web and mobile apps also differs. According to Maurer et al. [20], users prefer native apps over mobile websites, and likewise Serrano et al. [21] conclude that native apps provide a richer and more solid user experience.

Against this background, content and functionalities for mobile apps need to be designed differently in order to meet these specific challenges and create a good user experience. As a result, the methods for assessing the user experience in mobile app usage also need to be suitable for this context, in order to capture the mobile UX adequately. One of the challenges for mobile UX measuring is the mobile context of use "in the wild". We assume that not all methods for measuring UX can be directly applied to this context. And since there is no overview on methods for measuring mobile UX up-to-date, we conducted an initial literature survey that provides an overview of the current status quo of academic user experience evaluation methods for mobile applications on smartphones.

IV. Literature Survey

The literature survey focused on academic research papers listed in Google Scholar, as this is one of the most comprehensive and publisher-independent scientific literature databases available. As we are focusing on a smartphone-based understanding of mobile applications for our analysis, we excluded publications older than five years (before 2012) from our study. The UX evaluation methods we found in the literature were divided into three groups: (1) methods that have been directly applied from classical desktop UX scenarios to mobile applications without modification (*generic methods*); (2) methods that are adapted in order to be applicable to mobile applications (*mobile adapted methods*); (3) new approaches that have been developed for measuring UX specifically in the context of mobile applications on smartphones (*mobile specific methods*). For each of these three groups, we searched for methods of the five categories defined by Albert and Tullis [9]: Performance methods, usability methods, self-reporting methods, behavioral and physiological methods, and combined methods.

TABLE I. CATEGORIZED NUMBER OF UX STUDIES ON MOBILE APPS

	Generic Methods	*Mobile Adapted Methods*	*Mobile Specific Methods*
Performance Methods	0	0	2
Usability Methods	2	0	4
Self-Reporting Methods	4	1	0
Behavioral and Physiological Methods	0	0	6
Combined Methods	1	0	2
Sum	7	1	14

An overview on our results is displayed in Table I. In total, 22 studies were found. For a list of all studies that were used in the literature survey, see the Appendix. Though almost two thirds of the methods that we found –14 out of 22– were new approaches, there were seven methods that have been directly applied to the mobile context without being changed. There was only one approach that had been adapted to the context of mobile usage. In the following subsections, a qualitative description provides a deeper insight in the studies that were found.

A. Generic Methods

Whereas we found no example of performance methods that were directly applied or adapted to mobile apps, there are several usability studies that prove that issue-based methods work out fine for mobile apps as well. In [22], a classical usability lab study is conducted, supported by the use of two non-standardized questionnaires. Likewise, Habermann et al. [23] evaluated a public transportation app regarding its usability by observing users while they are solving prototypical tasks.

Dhir and Al-kahtani [24] used three standardized self-reported UX methods to evaluate a mobile app, i.e., the AttrakDiff questionnaire [25]. These methods have been directly applied to the mobile context without modification. Likewise, the standardized System Usability Scale (SUS) has been applied to various objects by Kortum and Bangor [26], but is also proven to be feasible for mobile apps in [27], using the questionnaire with ten mobile apps on smartphones as well as tablets while gaining meaningful results. Additionally, Ferreira et al. [28] used different self-reporting methods for mobile app UX evaluation: In the Expressing Experiences and Emotions (3E) method, as well as the Empathy Map (EM), users have to draw or write their feelings on a sheet of paper. Using Method of Assessment of EXperience (MAX), the participant has to sort cards on a board. In addition, the Self Assessment Manikin (SAM) questionnaire and Think Aloud were used. The methods were used to evaluate UX for apps on smartpones. However, all the studies were conducted in lab environments with no further consideration of the impact of a mobile usage situation. This might be the reason why none of them required particular adaptation for the mobile context.

As one example for a comprehensive combined approach, Yao et al. [29] conducted a mobile application user study in a lab setting, collecting task performance data, self-reported data, EEG data and skin conductance data. The results showed that these methods can be generically used in the context of mobile apps. Nevertheless, since all of them have been used in a lab study, their applicability depends on the mobility of the sensors that are used. We have no information about their operational capability in a real mobile context of use.

B. Mobile Adapted Methods

While several questionnaires and self-reported methods have been directly applied for mobile apps, there is also one study where an adapted method was used: The goal of Kujala and Miron-Shatz [30] is capturing the user experience from the actual context of smartphone use. In a long-term study, users have to fill out an initial questionnaire and two follow-up questionnaires after 2.5 and five months. They use a version of the Day Reconstruction Method (DRM) questionnaire, that is adapted to the context of the actual product use by adding new questions. In doing so, the users had to reconstruct and report the most important episodes of their day and their experiences and emotions during the episodes of smartphone use.

Apart from that, no studies on mobile adapted methods from the fields of performance, usability, behavioral and physiological methods, and combined methods could be found.

C. Mobile Specific Methods

Regarding the field of performance methods, there are some new approaches. Ravindranath et al. [31] introduce a new tool that monitors performance of mobile apps in the wild and helps diagnosing problems. Likewise, Liang et al. [32] developed a cloud service that traces mobile app performance and helps reducing crashes and performance bugs.

For measuring the usability of mobile apps, several new approaches have been introduced. Inspired by existing usability models, Harrison et al. [33] create People At the Center of Mobile Application Development (PACMAD), a model that conceptualizes usability particularly for the context of mobile applications. Igler et al. [34] show a framework that enables usability evaluation in the actual context of use instead of a lab setting. Olsina et al. [35] follow a novel holistic quality approach for the evaluation of usability and user experience of mobile applications. Furthermore, Hoehle and Venkatesh [36] developed a survey instrument based on a concept for mobile app usability that they derived from Apple's user experience guidelines.

In the field of behavioral and physiological methods, different mobile specific approaches have been developed. Yang et al. [37] use the front camera of smartphones to track the users' face expressions in order to enable facial aware applications. In [38], a software for mobile face tracking is presented, that analyzes emotions with an accuracy of 86 percent. Regarding brain activity measurement, there are some new approaches [39][40]. Both of them use a cap with EEG sensors that can be worn outside a lab scenario, while only Stopczynski et al. [39] combine it with open source software in order to visualize brain activity during smartphone use. However, both studies show the potential that lies within this method. Paletta et al. [41] and Kassner et al. [42] both used mobile eye tracking devices and their own software for collecting gaze data during smartphone usage. Paletta et al. [41] introduce the Smartphone Eye Tracking Toolbox (SMET), a software that i.a. records screencasts and collects data during an experiment. In their setting, eye tracking glasses are used, whereas Kassner et al. [42] developed a headset-like device for mobile eye tracking, that can not only be used for studies with mobile devices, but also for interaction with everyday objects.

In the field of combined methods, Maly et al. [43] introduce a new tool including usability measurement as well as skin conductance and heart rate in real mobile usage contexts. Participants are asked to walk along a pre-defined route in a building while assessing comprehensive data on their interactions, movements, stress level and physiological state. Noldus et al. [44] use both movement tracking and logging for automatically assessing mobile user experience. In their study, participants could move freely, while their movements were logged.

V. Discussion

The literature survey provided us with an overview of generic as well as mobile specific methods for evaluating the user experience of mobile applications. Though almost two thirds of the methods that we found were new mobile specific approaches, almost one third of the methods was directly applied to the mobile context without being changed. We found only one approach that had been adapted to the context of mobile usage.

The new approaches show that regarding performance measurement, evaluations of mobile apps often take place in lab settings and that for field studies with mobile devices, novel approaches are required. Though we identified several of these new methods, the topic still needs more coverage.

Regarding usability methods, the literature we found showed that usability methods have been directly applied to the mobile context, as far as they take place in a lab setting. For collecting usability information "in the wild", different and combined new approaches have been developed.

In the field of self-reporting methods, most of them have been directly applied to the mobile context without being modified, while there is no example for a new approach. One explanation may be that existing self-reported methods and questionnaires are standardized and seem sufficient for all contexts of use, since the only restriction for using them is the interaction with a system that is given with both desktop systems and mobile apps. In addition, questionnaires and reports can be filled in online and therefore are location-independent. Also, it lies within the nature of these methods that they are mostly carried out after the episode of interaction and therefore, the application exactly during the use of an app is not crucial for the use of these methods. One challenge in this field is constituted in those self-reporting methods that need moderation or guidance, since they require that both persons –moderator and participant– are in the same room, and often involve desk-bound actions like writing, drawing or sorting. To solve theses issues, new approaches for self-reported methods should be taken into account.

The behavioral and physiological methods that were found involved unexceptionally new, mobile specific approaches. The reason for that might be that all of the approaches in this topic are quite new, since the topic itself was only recently discovered in the context of user experience measurement and so far there are no established standards for this kind of methods. The fact that existing mobile hardware for measuring physiological conditions, such as caps with EEG sensors or eye tracking glasses, are fit for use in the mobile context, is clearly a chance for this field. On the other hand, these kind of measurements have to cope with a high amount of influences from the environment that makes the collected data hard to interpret. Thus, Maly et al. [43] point out that "*such an approach brings numerous methodological challenges as researchers do not control the environment setting and parameters [...]*". To our knowledge, to this point there is no solution to solve this issue.

Regarding the last field, we found only few studies with combined methods – one that could be directly applied to mobile context and two new approaches. The one that was directly applied to smartphone use was set in a controlled lab environment, so that under these circumstances, the mobile context is not given anymore. We therefore assume that there is a demand for comprehensive combined UX evaluation methods in the context of mobile applications.

VI. Conclusions and Further Research

In this paper, we have presented a literature survey on user experience evaluation methods for mobile applications on smartphones. We identfied academic research papers from five categories of UX methods and assigned them to three groups – generic methods, mobile adapted methods, and mobile specific approaches. As key findings, we can conclude that classic UX evaluation methods have been applied directly to the mobile context, as long as they are used in a lab setting, and that the main challenge of measuring mobile is data aquisition "in the wild", in the actual context of use. Based on our literature survey, we can furthermore identify several research needs for mobile UX research: (1) New approaches for self-reported methods should be taken into consideration; (2) new approaches for behavioral and physiological methods should be further developed towards standardized methods; (3) methods and frameworks for coping with physiological data collected via studies in uncontrolled environments are required.

Appendix

TABLE A.1. STUDIES USED IN THE LITERATURE SURVEY

Authors	*Classification*
Wei et al. (2015)	Generic methods - Usability
Habermann et al. (2016)	Generic methods - Usability
Dhir and Al-kahtani (2013)	Generic methods - Self-reporting
Kortum and Bangor (2013)	Generic methods - Self-reporting
Kortum and Sorber (2015)	Generic methods - Self-reporting
Ferreira et al. (2016)	Generic methods - Self-reporting
Yao et al. (2014)	Generic methods - Combined
Kujala and Miron-Shatz (2013)	Mobile adapted methods - Self-reporting
Ravindranath et al. (2012)	Mobile specific methods - Performance
Liang et al. (2014)	Mobile specific methods - Performance
Harrison et al. (2013)	Mobile specific methods - Usability
Igler et al. (2013)	Mobile specific methods - Usability
Olsina et al. (2014)	Mobile specific methods - Usability
Hoehle and Venkatesh (2015)	Mobile specific methods - Usability
Yang et al. (2012)	Mobile specific methods - Behavioral
Suk and Prabhakaran (2014)	Mobile specific methods - Behavioral
Stopczynski et al. (2014)	Mobile specific methods - Behavioral
Kranczioch et al. (2014)	Mobile specific methods - Behavioral
Paletta et al. (2014)	Mobile specific methods - Behavioral
Kassner et al. (2014)	Mobile specific methods - Behavioral
Maly et al. (2013)	Mobile specific methods - Combined
Noldus et al. (2014)	Mobile specific methods - Combined

Acknowledgment

This work was funded by the German Federal Ministry of Education and Research, grant no. 03FH032PX5; the PROFRAME project at RheinMain University of Applied Sciences. All responsibility for the content of this paper lies with the authors.

References

[1] AppAnnie, "The app economy forecast: $6 trillion in new value." [Online]. Available: https://www.appannie.com/de/insights/market-data/app-store-revenue-forecast-139-billion-2021/ 2017.08.30.

[2] AppAnnie, "Spotlight on consumer app usage: Part 2." [Online]. Available: https://www.appannie.com/de/insights/market-data/new-app-usage-report-how-many-apps-do-users-install-a-month/ 2017.08.30.

[3] V. Roto, E. L.-C. Law, A. P. O. S. Vermeeren, and J. Hoonhout, "User experience white paper: Bringing clarity to the concept of user experience." [Online]. Available: http://www.allaboutux.org/uxwhitepaper 2017.08.30

[4] V. Roto, "User experience definitions - all about ux," 2010. [Online]. Available: http://www.allaboutux.org/ux-definitions 2017.08.29.

[5] P. Morville, "User experience design," 2004. [Online]. Available: http://semanticstudios.com/user_experience_design/ 2017.08.30.

[6] M. Hassenzahl, "The hedonic/pragmatic model of user experience," Towards a UX Manifesto, vol. 10.

[7] J. J. Garrett, Elements of user experience, the: user-centered design for the web and beyond. Pearson Education, 2010.

[8] C. Stern, "Cubi: A user experience model for project success," 2014. [Online]. Available: https://uxmag.com/articles/cubi-a-user-experience-model-for-project-success 2017.08.29.

[9] W. Albert and T. Tullis, Measuring the user experience: collecting, analyzing, and presenting usability metrics. Newnes, 2013.

[10] J. Sauro and J. R. Lewis, Quantifying the user experience: Practical statistics for user research. Morgan Kaufmann, 2016.

[11] A. P. O. S. Vermeeren et al., "User experience evaluation methods," in Proceedings of the 6th Nordic Conference on Human-Computer Interaction, E. T. Hvannberg, Ed. New York, NY: ACM, 2010, pp. 521–530.

[12] E. L.-C. Law, P. van Schaik, and V. Roto, "Attitudes towards user experience (ux) measurement," International Journal of Human-Computer Studies, vol. 72, no. 6, 2014, pp. 526–541.

[13] S. Hart, "Mobile vs. desktop: 10 key differences," 2017. [Online]. Available: https://www.paradoxlabs.com/blog/mobile-vs-desktop-10-key-differences/ 2017.08.31.

[14] J. Nielsen and R. Budiu, Mobile usability. MITP-Verlags GmbH & Co. KG, 2013.

[15] M. Firtman, Programming the mobile web. " O'Reilly Media, Inc.", 2010.

[16] C. P. Furner, P. Racherla, and J. S. Babb, "Mobile app stickiness (mass) and mobile interactivity: A conceptual model," The Marketing Review, vol. 14, no. 2, 2014, pp. 163–188.

[17] C. Furner, P. Racherla, and J. Babb, "What we know and do not know about mobile app usage and stickiness: A research agenda," Geospatial Research: Concepts, Methodologies, Tools, and Applications, vol. 1, 2016, pp. 117–141.

[18] A. Mendoza, Mobile user experience: Patterns to make sense of it all. Waltham, MA: Morgan Kaufmann, 2014.

[19] A.-L. Selke, "Desktop vs mobile: Do devices change our behaviour?" 2015. [Online]. Available: https://www.nathalienahai.com/2015/11/desktop-vs-mobile-do-devices-change-our-behaviour/ 2017.08.31.

[20] M.-E. Maurer, D. Hausen, A. De Luca, and H. Hussmann, "Mobile or desktop websites?: Website usage on multitouch devices," in Proceedings of the 6th Nordic Conference on Human-Computer Interaction: Extending Boundaries, ser. NordiCHI '10. New York, NY, USA: ACM, 2010, pp. 739–742.

[21] N. Serrano, J. Hernantes, and G. Gallardo, "Mobile web apps," IEEE software, vol. 30, no. 5, 2013, pp. 22–27.

[22] Q. Wei, Z. Chang, and Q. Cheng, "Usability study of the mobile library app: An example from chongqing university," Library Hi Tech, vol. 33, no. 3, 2015, pp. 340–355.

[23] A. L. Habermann, K. Kasugai, and M. Ziefle, "Mobile app for public transport: A usability and user experience perspective," in Internet of Things. IoT Infrastructures, ser. Lecture Notes of the Institute for Computer Sciences, Social Informatics and Telecommunications Engineering, B. Mandler, J. Marquez-Barja, and M. E. Mitre Campista, Eds., 2016.

[24] A. Dhir and M. Al-kahtani, "A case study on user experience (ux) evaluation of mobile augmented reality prototypes," J. UCS, vol. 19, no. 8, 2013, pp. 1175–1196.

[25] User Interface Design GmbH, "AttrakDiff," 2017. [Online]. Available: http://attrakdiff.de/index-en.html 2017.09.28.

[26] P. T. Kortum and A. Bangor, "Usability ratings for everyday products measured with the system usability scale," International Journal of Human-Computer Interaction, vol. 29, no. 2, 2013, pp. 67–76.

[27] P. Kortum and M. Sorber, "Measuring the usability of mobile applications for phones and tablets," International Journal of Human-Computer Interaction, vol. 31, no. 8, 2015, pp. 518–529.

[28] B. M. Ferreira et al., "Evaluation of ux methods: Lessons learned when evaluating a multi-user mobile application," in International Conference on Human-Computer Interaction. Springer, 2016, pp. 279–290.

[29] L. Yao et al., "Using physiological measures to evaluate user experience of mobile applications," in International Conference on Engineering Psychology and Cognitive Ergonomics. Springer, 2014, pp. 301–310.

[30] S. Kujala and T. Miron-Shatz, "Emotions, experiences and usability in real-life mobile phone use," in CHI 2013, W. E. Mackay, S. Brewster, and S. Bødker, Eds. New York, NY: ACM, 2013, p. 1061.

[31] L. Ravindranath et al., "Appinsight: Mobile app performance monitoring in the wild." in OSDI, vol. 12, 2012, pp. 107–120.

[32] C.-J. M. Liang et al., "Caiipa," in Proceedings of the 20th International Conference on Mobile Computing and Networking, September 7 - 11, 2014, Maui, HI, USA, S.-J. Lee, Ed. New York, NY: ACM, 2014, pp. 519–530.

[33] R. Harrison, D. Flood, and D. Duce, "Usability of mobile applications: Literature review and rationale for a new usability model," Journal of Interaction Science, vol. 1, no. 1, 2013, pp. 1–42.

[34] B. Igler, T. Braumann, and S. Böhm, "Evaluating the usability of mobile applications without affecting the user and the usage context," International Journal Of Business and Management Studies, vol. 5, no. 1, 2013, pp. 93–102.

[35] L. Olsina, L. Santos, and P. Lew, "Evaluating mobile app usability: A holistic quality approach," in Web engineering, ser. Lecture Notes in Computer Science, S. Casteleyn, G. Rossi, and M. Winckler, Eds. Cham: Springer, 2014, vol. 8541, pp. 111–129.

[36] H. Hoehle and V. Venkatesh, "Mobile application usability: Conceptualization and instrument development," Mis Quarterly, vol. 39, no. 2, 2015, pp. 435–472.

[37] X. Yang et al., "Visage: A face interpretation engine for smartphone applications," in International conference on mobile computing, applications, and services. Springer, 2012, pp. 149–168.

[38] M. Suk and B. Prabhakaran, "Real-time mobile facial expression recognition system - a case study," in Proceedings of the IEEE Conference on Computer Vision and Pattern Recognition Workshops, 2014.

[39] A. Stopczynski et al., "Smartphones as pocketable labs: Visions for mobile brain imaging and neurofeedback," International journal of psychophysiology, vol. 91, no. 1, 2014, pp. 54–66.

[40] C. Kranczioch, C. Zich, I. Schierholz, and A. Sterr, "Mobile eeg and its potential to promote the theory and application of imagery-based motor rehabilitation," International journal of psychophysiology, vol. 91, no. 1, 2014, pp. 10–15.

[41] L. Paletta et al., "Smartphone eye tracking toolbox: Accurate gaze recovery on mobile displays," in Proceedings of the Symposium on Eye Tracking Research and Applications, ser. ETRA '14. New York, NY, USA: ACM, 2014, pp. 367–68.

[42] M. Kassner, W. Patera, and A. Bulling, "Pupil: An open source platform for pervasive eye tracking and mobile gaze-based interaction," in Proceedings of the 2014 ACM International Joint Conference on Pervasive and Ubiquitous Computing: Adjunct Publication, ser. UbiComp '14 Adjunct. New York, NY, USA: ACM, 2014, pp. 1151–1160.

[43] I. Maly, Z. Mikovec, J. Vystrcil, J. Franc, and P. Slavik, "An evaluation tool for research of user behavior in a realistic mobile environment," Personal and Ubiquitous Computing, vol. 17, no. 1, Jan 2013, pp. 3–14.

[44] L. P. Noldus, M. K. Ben Loke, and A. J. Spink, "Automated mobile user experience measurement: Combining movement tracking with app usage logging," Creating the Difference, vol. 31, 2014.

Hedonic Motivation of Chatbot Usage

Wizard-of-Oz Study based on Face Analysis and User Self-Assessment

Judith Eisser and Stephan Böhm

CAEBUS Center of Advanced E-Business Studies
RheinMain University of Applied Sciences
Wiesbaden, Germany
Email: {judith.eisser, stephan.boehm}@hs-rm.de

***Abstract*—Ever since, the Internet enabled conversations by supporting interactive features and the presentation of product- or service-related information accumulated on websites, micro pages or portals. More recently, especially user groups of younger generations turn towards messaging applications for communication. Companies adjust to this trend of more messaging-oriented forms of interaction by implementing new channels of customer communication such as chatbots. In this work, a comparative analysis is conducted to uncover the impact of using traditional websites or chatbots for promoting a product in an impulse purchase situation with special attention to hedonic motivation. The aim is to measure the impact of the information delivery option (website or chatbot) on the customers' emotions as expression of hedonic motivation. More specifically, this paper is addressed to answering the question whether chatbot utilization result in a different hedonic motivation and in turn a higher manifestation of positive emotions than tradition website usage. The chatbot-based scenario is implemented by using a Wizard-of-Oz (WOz) experimental approach. The results provide first insights on the effects of chatbot usage on emotions in electronic commerce environments: while the chatbot users showed slightly higher happiness scores, no statistically significant impact could be discovered and there does not seem to be a statistically significant influence of chatbot usage on the purchase decision.**

Keywords–Chatbots; Conversational Commerce; Comparative Analysis; Wizard-of-Oz; Emotion Recognition; Hedonic Motivation.

I. INTRODUCTION

In recent years, online communication has shifted from a one-way to a conversational approach [1]. Internet users do not only receive information but also generate content themselves and interact in networks, e.g., via online communication applications. In this network environment, information is not only pushed, but also actively pulled by the users suited to their specific requirements [2]. In doing so, they expect to be treated individually and to be singularly addressed while their questions are adequately answered [3]. Hence, it can be said that the common website behavior of searching and finding transforms into a process of asking and receiving answers. Companies are adapting to this transformation by increasingly offering and sharing information, as well as promotions in online channels allowing for two-way conversations. A current trend is the implementation of chatbots. According to Mittal et al., chatbots are conversational programs for question-and-answer processes, which interact with the users in the form of ever-present assistants. Such systems can be based on pattern matching and natural language processing methods or artificial intelligence (AI) techniques [4]. These days, big global players, such as Google and Microsoft are conducting extensive research in order to advance this technology [5], which shows the current interest and importance of the technology.

Within e-business, chatbots can not only be used for focused product inquiries but also for product comparison or to assist users within the product decision-making processes [6]. Currently, more than a hundred thousand unique chatbots are being offered (with the Facebook messenger being the most popular implementation platform) [7]. As reported in [8], more than one fifth of the population in the US has used such chatbots offerings already. According to a study by Oracle and Coleman Parks, 80 percent of the queried companies either already have implemented or plan to implement a chatbot into their marketing strategy in order to improve their customer experience by 2020 [9]. Thus, the relevancy of the topic becomes apparent.

In ligth of the above, this article is about the assessment of hedonic motivation within impulse purchase situations focusing on chatbot and website utilization. It proceeds with a research background where conversational commerce and chatbot utilization are examined as well as the according role of hedonic motivation. In Section III, our approach on measuring hedonic motivation based on a Wizard-of-Oz testing scenario is presented. Section IV contains details concerning the experimental study while the results are presented in Section V. There, we discuss general findings, the differences between emotion self-assessment and face analysis, as well as the analysis concerning hedonic motivation. In the last section, a conclusion is given concerning the study at hand.

II. RESEARCH BACKGROUND

The analysis of the hedonic motivation of chatbot usage requires a more comprehensive understanding of the principles of the emerging conversational commerce and the concept of hedonic motivation. For this reason, we will provide a more detailed background on the related fundamentals and related scientifc work for these two topics in the following.

A. Chatbot Usage and Conversational Commerce

Chatbots are dialogue programs in the form of composed pattern matching and natural language processes or artificial intelligence techniques [4], which can effectively be used for interactive question answer processes [10]. There are early examples of such systems that date back into the mid-1960s such as the popular ELIZA system and many more systems that have been discussed in literature (e.g., Albert One, ALICE)

[11]. Pattern matching or rule-based processing of chatbots are searches for key words, word roots and synonyms for example. They are noted in code in order to predefine possible conversation flows to generate answers to questions [11]–[13]. In this so-called retrieval-based approach, the chatbot produces answers from a predefined database according to rules. Artificial intelligence techniques in the form of generative models go beyond this logic of predefinition by allowing for learning processes where the bot program generates unique answers via knowledge assembly and by analyzing the current context [14]. Tailored to interaction with humans –consumers in this case–, they produce and understand written input in natural language. This human interface offers an interesting alternative compared to the traditional information architecture, where information is structured and formatted to interact with screens. Without particular setup requirements, it can be easily interacted with and utilized [15]. Application fields for chatbots can be educational, customer service or e-commerce scenarios for example [16].

Within e-commerce, one trend gaining interest is conversational commerce [17]. Chris Messina, chatbot industry expert and trend watcher, created the industry-wide accepted definition (e.g., [18][19]) for conversational commerce, which is about "utilizing chat, messaging, or other natural language interfaces (i.e., voice) to interact with people, brands, or services and bots that heretofore have had no real place in the bidirectional, asynchronous messaging context" [20]. Commercial chatbot conversation can be seen as a part of conversational commerce, as the latter can be seen as a combination of messaging apps or rather human-bot chatting and shopping in the form of conversational customer interaction [17]. As such, chatbots transport the previously mentioned idea of asking and and receiving answers into e-commerce allowing consumers to naturally engage with companies in a commercial context like they are used to through common interpersonal conversation. This rather natural engagement capability can be seen as one of the main advantages of conversational commerce alongside the easy accessibility and the already familiar interface within messaging apps [17]. Utilized in such an e-commerce context, chatbots can also improve customer satisfaction [21].

B. Hedonic Motivation in Conversational Commerce

Hedonic shopping can be operationalized through several items, such as joy, excitement, arousal, festive, escapism, fantasy, and adventure, as stated for example in [22]–[26]. A value for the total hedonic motivation can be calculated as the sum of positive and negative emotions [27]. This value is a key element of the consumer experience [28][29]. Emotions are a significant part of the Stimulus-Organism-Response (SOR) process [30][31], which is used in marketing science or to measure hedonic benefits [32][33]. The aim of the consumer is to select hedonic experiences, which maximize positive emotions [34]. The emotional aspect of hedonism can be seen as essential for user satisfaction in information systems [35]. Defined as wish to satisfy a need, hedonic motivation can be specified as emotional experience [31][36], which has been found to directly influence the consumers' positive emotional responses [31]. Further studies examined the aspects of emotions within chatbot usage in general (e.g., [37][38]) or the influence of hedonic values for e-service quality [39] for example. To the authors knowledge, there is no existing study combining these aspects by analyzing emotions and chatbot utilization in an e-shopping situation. In our study, the multidimensional construct emotion is operationalized solely as a physically measurable expression based on the seven basic universal emotions as discovered by Ekman and Heider [40]: Anger, fear, sadness, disgust, surprise, contempt and happiness. Happiness, as emotional aspect, will be focused on in our study. This is because, as per definition, within hedonic motivation positive emotions shall be maximized from consumer side and thus represent an important part of the consumers' hedonic motivation [31].

Impulse purchasing can be seen as hedonic purchase behavior and consumer emotions have been found to influence impulse buying behavior – happiness or excitement being a positive influence [31]. Hence, this study examines the impact of chatbot usage in an impulse purchase situation. For this purpose, a comparative study based on a traditional webpage-based e-shopping scenario and a chatbot enhanced variant is used to analyze for differences in the resulting customers' hedonic motivation.

Different digital offers such as social network sites, social media in general, shopping sites or dialogue systems inherit different content and are set up differently as well thus can be examined individually concerning the influence of hedonic aspects (e.g., [35][41]). In light of the current trend of implementing conversational offers into commerce contexts, we expect diverging manifestations of happiness. Thus, we think that there are different levels of hedonic motivation when consumers interact solely with product webpages or get assistance from chatbots. While previous research already integrated hedonic motivation into impulse buying behavior research and chatbot research in the form of chatbot metrics frameworks (e.g., [31][42]), there is a research gap concerning the combination of these aspects, which this study aims to bridge. Hence, the purpose of this study is to explore the differences in hedonic motivation in the form of happiness scores between traditional product websites and chatbot enhanced customer interaction in an impulse purchase situation. More specifically, the aim of our experimental study is to empirically validate the following research questions:

1) Which levels of hedonic motivation do consumers have when interacting with a chatbot or browsing a website for product information?
2) Can happiness as an operationalization of hedonic motivation be consistently measured by face analysis and self-assessment?
3) To what extent does the use of chatbots as a way to enhance product websites have a measurable impact on hedonic motivation in impulse purchase situations?

III. APPROACH

The study at hand uses a Wizard-of-Oz testing approach to simulate the integration of an advanced chatbot system on an e-commerce website. Emotional self-assessment and a face analysis software for video files are used to measure the user's state of happiness in our experimental scenario. Before we describe the configuration of the study in more detail, some methodological details of our approach will be described in this section.

A. *Wizard-of-Oz (WOz) Testing in Chatbot Research*

The WOz method does not only represent a way to investigate immature technology in a prototypical manner but it is also a way to avoid prohibitively high costs and time efforts [43] and to enable a testing environment without the need of coding is the utilization of a Wizard-of-Oz approach. The approach is defined as a kind of simulation where researchers "conceal themselves from research participants and use communications technology to pretend that a prototype or incomplete computer-based conversational system is fully functioning" [43].

In order to conduct a WOz-study, several aspects can be taken into consideration according to Eynon et al. [43]:

- Prototype functionality and fidelity of the prototype,
- technical handling of the prototype by the wizard,
- wizard visibility and control,
- user knowledge concerning the WOz setup,
- research design (Controlled experiment vs. uncontrolled exploration).

Wizard-of-Oz dialogues as utilized for chatbot examinations hold the advantage of resembling realistic behavior, which can appear to be more capable than already existing dialogue systems [6]. They can be seen as a feasible way to cope with the lack of technological advantage in order to assess chatbots as a suitable way of product promotion [10]. They become relevant because rule-based Chatbot dialogue systems as developed until today are limited in functionality since they are bound to a pre-defined database [14] and not able to learn. AI-based chatbot solutions are in an early stage [14] and the implementation of such systems can be a challenging task requiring comprehensive technical knowledge. Hence, this study made use of a WOz approach to simulate an advanced chatbot solution without the need for a complex technical implementation.

Commonly used in chatbot research, for example when setting up interactive question answering systems in the form of a chatbot [10], when examining the role of memory in goal-oriented dialogue systems [6] and when studying non-verbal processing in general [44], the method is being utilized for the study at hand as well. According to the components by Eynon et al., the prototype is set up as a fully functional web-based chatbot with a trained wizard controlling the prototype, who is hidden from the participants in an environment of tight experiment control [43].

B. *Measuring Hedonic Motivation*

Two methods of hedonic motivation measurement are utilized in this study: emotion self-assessment by the participants and face analysis via a cloud-based face analysis tool. Both practices are being explained in the following.

1) Emotion Self-Assessment: Hedonic motivation can be measured on the basis of different aspects such as the levels of excitement, arousal or escapism (e.g., [22]–[26]). For the study at hand, emotions and within this construct happiness in particular is the relevant aspect to consider. One way of assessing emotions within a research study is the distribution of suitable questionnaires before, during and/or after the session [45]. This written-down method of emotion self-assessment is applied in this study in order to control for possible mismatches

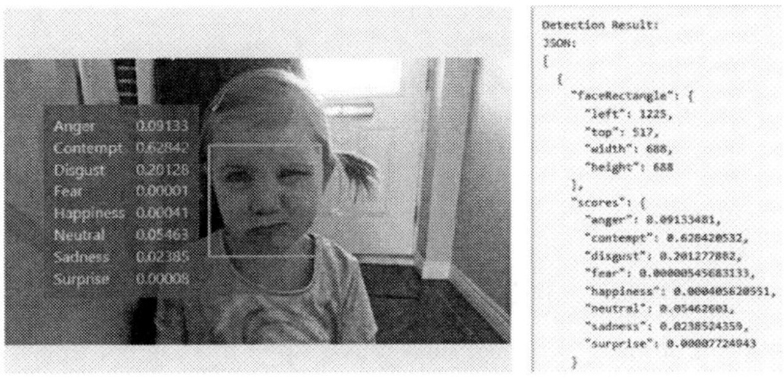

Figure 1. Exemplary Microsoft Emotion API face analysis result. Reprinted from Kearn [49]. Copyright by Microsoft [2016].

in comparison to the face analysis results – the participants conducted it prior to the session and immediately after pre-defined stimuli or impulses. The individual manifestations of the basic emotions as defined by Ekman and Heider [40] were requested via a ten-stage rating scale before converting them to values between zero and one. This was necessary in order to be consistent with the value range of the emotion data derived from the face analysis tool.

2) Emotion Tracking and Face Analysis: On a physical level, the emotions as the individual manifestations of the different emotional values according to Ekman and Heider [40], can be assessed via face analysis. On basis of this emotion classification, the face muscles and the according mimic manifestations are being analyzed and categorized into the seven distinct emotions [46]. Such an analysis can be conducted based on video material of the participants' faces recorded during experimental sessions where individuals are exposed to defined impulses to stimulate the expected reaction or emotion. Face analysis requires complex algorithms and massive data processing but is also available by cloud-based services like the Microsoft Cognitive Services. The service used in this paper is the Microsoft Emotion API, which provides emotional scores based on the seven basic emotions fear, anger, sadness, disgust, surprise, contempt and happiness. Additionally, the service calculates a neutral score indicating an absence of the other seven emotions within the measurement out of face recordings of the participants. The scores are to be interpreted as normalized scores ranging from zero to one – thus, the program shows the relative scores of the eight different emotional states indicating the predominant ones. This information is not to be confused with emotion intensity, which no information is given for within the tool results [47]. According to Microsoft, the two emotions contempt and disgust are only experimental for now [48]. Since neutral can be seen as the absence of the other emotional scores, the focus will be on those other scores rather than on neutral.

IV. Experimental Study

Based on the theoretical and methodological foundation above, an experimental study on the hedonic motivation of chatbot usage is presented in this section. The study is a work-in-progress and was based on an experimental setup with a convenience sample. Besides validating the research questions presented in Section II, the aim of the study is to derive some more insights on the applicability of the discussed test setups and tools for chatbot prototyping and emotion measurement.

A. *Wizard-of-Oz Setup*

Starting point of the experimental study was the development of a chatbot concept. The utilized chatbot prototype for our WOz study has been designed as a web-based bot. This means that it searches for key words in the form of full words, their lemmas, roots and synonyms. The according information needs to be fed into the chatbots database prior to utilization. Based on this data and the defined rules, the chatbot responds with a preset answer. A fallback answer has been defined to handle questions new to the chatbot to ensure consistency concerning the user experience ("Unfortunately, I did not entirely understand you. How exactly can I help you?"). This inquisitive aspect can be built in in case of insufficient information in order to be able to give an adequate answer [12]. Such a response does not only inform the user that the system cannot process his question but also prompts the user to rephrase or state his question more clearly. Figure 2 shows an exemplary chat snippet from the study, which has been translated from German into English language.

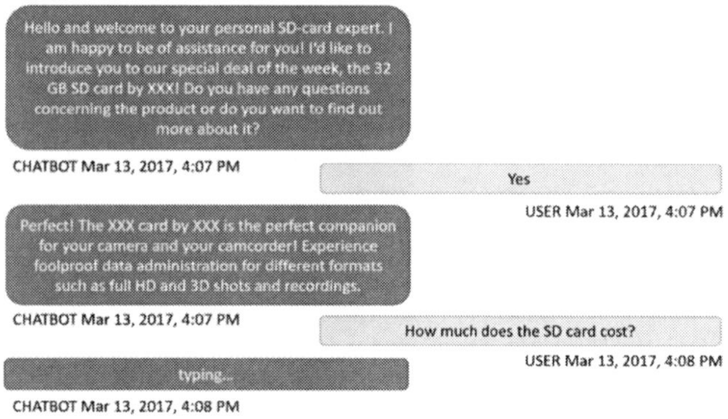

Figure 2. Exemplary chatbot dialogue snippet

For reasons of simplification, our concept does not adapt its behavior and answers based on previous customer input. This means that there are no adaptable templates defined but a static set of predefined responses. Based on the concept above, the chatbot was set up with the Wizard-of-Oz method, where no program converses with the user. The response is controlled by a human counterpart —the wizard—, who applies pre-set rules implemented prior to application. The WOz setup of this study is an own web solution running via the chatserver Arrowchat [50] filled with predefined JavaScript commands, which are elicited by a human via a control board. The wizard can respond to the participants interaction by selecting pre-defined phrases from this control board. According to these commands, answer sentences are being shown to the participant through a chat interface. The wizard could also type in text as a fallback feature if the set of predefined phrases is not sufficient. However, this option was only used as an exception to cope with unpredicted user behavior.

The experimental setup of the study is shown in Figure 3. Two separated rooms were used. The survey, briefing of the participant on the course of the study, as well as an interview were conducted in room 1. After the study preparation the wizard left room 1 to operate the control panel and supervise the experiment from room 2. The participants were left unattended to complete the defined task the chatbot system. They were not informed to participate in a WOz scenario and did not know that they interact with the wizard in room 2. Two

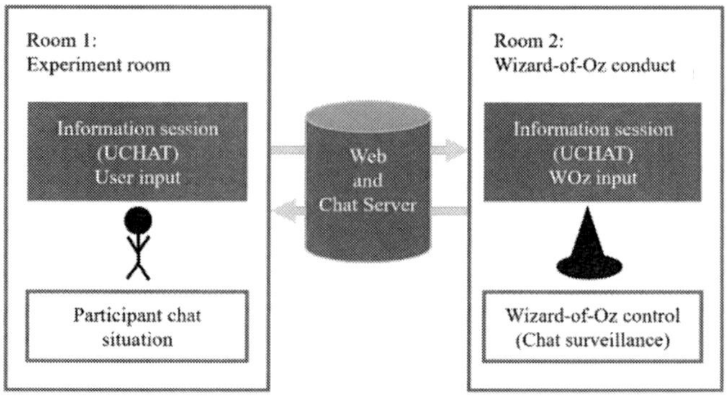

Figure 3. Wizard-of-Oz experiment setup

laptops have been utilized – all participants were positioned in front of the first to be recorded and to interact with the chatbot or the product webpage under the same conditions. The second laptop has been used by one of the authors to act as the chatbot wizard.

B. *Study Procedure*

For the laboratory situation, the participants were randomly assigned to conduct a product information process either via chatbot (UCHAT) or via a product information detail page (UPAGE). The sessions were divided into five procedural steps:

1) Introduction to the study with its purpose and signing of the consent form concerning the recording and data processing method via the utilized cloud-based face analysis tool.
2) Questionnaire with socio-demographic aspects such as age, job position and education as well as personality traits according to the Big Five approach (the five dimensions of personality: openness, conscientiousness, extraversion, agreeableness and neuroticism) by Rammstedt et al. [51].
3) Screen capturing and emotion tracking while accomplishing the task of browsing a product overview website for a specific product (digital cameras).
4) Simulation of an impulse purchase situation by displaying a product promotion in a pop-up window on the webpage.
5) Structured interview with detailed questions concerning the experience during the session.

Steps two to four were time-constraint by preset appearance times set on the utilized laptop for comparability. The simulation of the impulse purchase situation by the pop-up windows consisted of two defined stimuli or impulses that initiated and ended this phase of the experiment:

- Impulse 1 (I1): Overlay of a specific product promotion (SD card) via a pop-up or product overview website five seconds after accessing the site. Showing an instruction to open the product detail page or the chatbot respectively according to their affiliation (UCHAT or UPAGE).
- Impulse 2 (I2): Exposure to a purchase solicitation with question concerning the purchase decision. Showing an instruction to answer to the according question of the chatbot (UCHAT) or click accordingly within the appearing pop-up (UPAGE).

The participants were not informed about the precise time constraints in order to simulate a natural product information process situation. Since neutral can be seen as the absence of the other emotional scores, the emotion assessment focus was on those other scores rather than on neutral.

For the analysis of the data, videos, audio files as well as questionnaire responses, several steps were necessary: The video preparation consisted in a (1) systematic filing of the two videos per participant (screen capturing and face recording), (2) the cutting according to the two impulses I1 and I2, thus creation of two 20-second videos per testing person, (3) the appropriate formatting and labeling of the video files for the cloud-based analysis, (4) the retrieval of the emotion data by processing the video files through the Microsoft Emotion API and (5) the transferral of the raw analysis data into the data processing program and statistical analysis. For the audio interviews, the data was transcripted and analyzed. Alongside with the quantitative survey data, the material was statistically analyzed and descriptively evaluated as stated in the following sections.

C. Sample Description

Among the 57 participants, the chatbot was used by 28 (= UCHAT group) while 29 formed the control (UPAGE) group by using the unassisted product webpage only. Table I gives an overview of the demographics of the sample. The sum of the figures does not always add up to 57 because of missing values in the datasets.

TABLE I. DEMOGRAPHIC SAMPLE DESCRIPTION

		Total	*UCHAT*	*UPAGE*
Gender	Female	39	38	21
	Male	16	9	7
Mean age (in years)		25.4	24.6	26.1
Educational level	Student	35	19	16
	Pupil	5	1	4
	Other	15	7	8
Highest educational degree	University level	15	7	8
	A-Level	37	17	20
	High School	3	1	2
Chatbot experience	No	37	14	23
	Yes	19	14	6

Most of the participants are female, in their mid-twenties, A-level (university entry qualification) university students and do not have prior chatbot experiences yet. This is due to the fact that this experiment was conducted in an university environment based on a convenience sample. This has to be considered with regard to the generalizability of the findings. However, as this pre-study focuses more on the general measurability of effects and the applicability of the suggested approach, the composition of the groups is not an issue here.

V. RESULTS

The experimental study generated data that required further analysis to answer the research questions and aims that have been defined above. Before the according results on emotion measurement and chatbot impact are presented, we introduce this section with some more general findings for the UCHAT and UPAGE groups in the sample.

A. General findings

During the session, much information has been obtained concerning the participants' experience with online shopping in general (frequency of online product information searches and purchases) as well as with chatbot dialogues, their opinion on the product search conducted during the session (purchase decision, influencing factors on the decision, their feeling during the conduct, their own preferred way of informing themselves online) and their emotional states prior to the session and after impulses I1 and I2.

Table II shows the manifestations of the aspects stated above for the two groups UCHAT and UPAGE. It can be seen that most are experienced and active e-commerce users. Most of the users in both groups are using the Internet for searching product information more than once a week. The participants are even more active in purchasing online as the majority in both groups purchases products online at least one a week.

TABLE II. UCHAT AND UPAGE ONLINE EXPERIENCE

Aspect		*UPAGE*	*UCHAT*
Frequency of online product search	Fewer than once a month	5	3
	1-2 times a month	11	4
	At least once a week	8	7
	Daily	5	3
Frequency of online purchases	Fewer than once a month	0	2
	1-2 times a month	8	5
	At least once a week	18	18
	Daily	3	1

Missing values occur due to not mentioned aspects within the structured interviews at the end of the session. It can be seen that according to the Spearman correlation analysis, the two differences in online experience between the two groups are not significant ($p = .179$ for the frequency of online information and $p = .864$ for the frequency of online purchases) meaning that both groups did not differ concerning online expertise.

Other interesting aspects seemingly diverging are presented in Table III. It can be seen that differences occured between the two groups with regards to chatbot experience, the perceived feeling of being well-informed, having missed information within the information process, the influence of the received information on the purchase decision and the individual purchase decision.

TABLE III. SELECTED UCHAT AND UPAGE EXPERIMENT RESULTS

Aspect		*UPAGE*	*UCHAT*
Existing chatbot experience	No	23	14
	Yes	6	13
Feeling of being well-informed	No	3	6
	Yes	26	19
Feeling of having missed information	No	23	15
	Yes	6	12
Feeling of a good experience	No	13	7
	Yes	12	19
Positive purchase decision	No	20	16
	Yes	9	12
Decision influenced by information	No	14	9
	Yes	12	12

The differences in feeling of having missed information and the influence of the given information on the purchase decision seem ample – however, none of the differences are statistically significant (Pearson Chi square p-values higher than 0.01).

B. Self-Assessment and Face Analysis

The face analysis of the video material generated time-coded emotion values. We focused the data analysis on 10 seconds before and after the defined impulses I1 and I2. Figure 4 shows an exemplary data flow from the Microsoft Emotion API tool. In this study, this data is compared to the self assessed emotional manifestations of the participants.

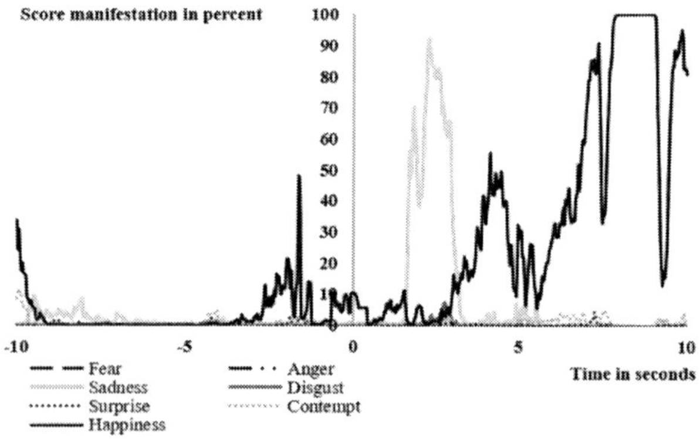

Figure 4. Exemplary happiness score flow around impulse I1

The scores for the happiness value were extracted from the data and mean values for the scores have been calculated to represent the observation time frame for each participant and the respective two stimuli. This data is then compared to the self-assessed emotion manifestations. Table IV shows the specific happiness scores the participants possessed during I1 and I2 of the online face analysis tool and the self-assessment in order to assess the differences the tool measured and perceived hedonic motivation of the participants.

TABLE IV. HAPPINESS SCORES VIA FACE ANALYSIS (FA) AND SELF-ASSESSMENT (SA)

Stimulus		*FA*	*SA*
Impulse 1 (I1)	Mean	0.234	0.272
	SD	0.353	1.698
Impulse 2 (I2)	Mean	0.278	0.354
	SD	0.382	1.984

The happiness score means presented in table IV are shown as calculated means over the whole timespan across either I1 or I2 and are displayed as values between 0 and 1. SD, the standard deviation, is displayed as percentage points. Both kinds of assessment show significant differences (p lower than .001 for both I1 and I2). Thus, the results of online face analysis and emotion self-assessment diverge. This is a preliminary result and needs more investigation. However, it might give some indication that the self-assessed emotion cannot be represented by a mean value of an emotional status measured over time, additional factors forming the emotional status after the stimulus or different value interpretations (predominance vs. intensity).

C. Hedonic Motivation Analysis

Table V shows the happiness scores the participants possessed during I1 and I2 in order to assess the differences in hedonic motivation of the participants concerning their affiliation to the UCHAT and UPAGE groups. The self-assessed scores have been taken for analysis because of the diverging results as discovered in sub section B and the resulting decision to focus on one of the two methods. When comparing the mean values, the happiness scores are slightly higher in the group of the sample that was assisted by the chatbot. This could indicate that the usage of chatbots has a positive impact on customers' hedonic motivation. However, a more comprehensive analysis reveals, that the difference is not statistically significant ($p = .55$ for I1, $p = .148$ for I2).

TABLE V. HAPPINESS SCORES OF UCHAT AND UPAGE

Stimulus		*UCHAT*	*UPAGE*
Impulse 1 (I1)	Mean	0.286	0.258
	SD	0.146	0.192
Impulse 2 (I2)	Mean	0.314	0.238
	SD	0.184	0.208

In Table VI, the happiness scores for the groups of buyers and non-buyers can be seen. From a purely descriptive perspective it is interesting, that we could observe (1) a slightly higher increase of happiness within the group of buyers and (2) a higher mean score for happiness of the buyers compared to the non-buyers for I2 (purchase question).

TABLE VI. HAPPINESS SCORES OF BUYERS AND NON-BUYERS

Stimulus		*Buyer*	*Non-buyer*
Impulse 1 (I1)	Mean	0.333	0.236
	SD	0.185	0.151
Impulse 2 (I2)	Mean	0.338	0.239
	SD	0.201	0.190

However, also for the purchase decision there is no significant difference concerning the means when statistically analyzed ($p = .036$ for I1 and $p = .068$ for I2).

VI. CONCLUSION AND FUTURE WORK

The study applied WOz testing to analyze the impact of chatbot usage on hedonic motivation. Data on the emotional status of the participants was acquired based on face analysis and self-assessment. Levels of hedonic motivation could be measured but the data revealed no statistically significant differences for users with and without chatbot support. Moreover, there was no statistical significant difference between those groups with regard to the purchase decision in a simulated impulse purchase situation.

This does not mean that chatbots do not have an impact on purchase behavior as our observations depend very much on the study sample and setup. However, our results could give some first indication, that the value add generated by implementing chatbots must address aspects beyond pure enjoyment or producing "happier" customers. Another interesting finding of the study is that measures of happiness as operationalization of hedonic motivation by face analysis and self-assessment did not produce consistent results. As mentioned before, the reasons can be manifold. Obviously, there is a difference between the analysis of facial expressions within a specific time-frame and self-assessments that are based on perceptions and experiences. Additionally, commercially available tools are somehow "black boxes" with regard to the algorithms involved or the calculation and interpretation of resulting scores. Thus, researchers must be careful when using an appropriate approach for emotion detection and measurement.

The study exhibits several limitations. The results may not be generalizable as the study was conducted based on a convenience sample as mentioned before. Furthermore, the impulse purchase situation was simulated with preselected products and defined impulses, which might have influenced the participants' emotional states. Their opinion of the product might have overshadowed the potential impact they might have been exposed to when being assigned to the UCHAT or UPAGE group. Being limited to the emotional state of happiness as representative of the aspect of joy as defined by [22] for example, future research might investigate other aspects of hedonic motivation in online shopping such as arousal or escapism.

REFERENCES

[1] J. Rowley, "Just another channel? Marketing communication in e-business," Marketing Intelligence & Planning, vol. 17, no. 4, 2006, pp. 332–351.

[2] C. Gurau, "Integrated online marketing communication: implementation and Management," Journal of Communication Management, vol. 12, no. 2, 2008, pp. 169–184.

[3] S. Holtz, Public Relations on the Net: Winning Strategies to Inform and Influence the Media, the Investment Community, the Government, the Public, and More! New York: AMACOM, 2002.

[4] A. Mittal, A. Agrawal, A. Chouksey, R. Shriwas, and S. Agrawal, "A comparative study of chatbots and humans," International Journal of Advanced Research in Computer and Communication Engineering, vol. 6, no. 6, 2016, pp. 1055–1057.

[5] J. Günther, "Digitalisation in purchasing - technological and organizational trends," Elektrotechnik & Informationstechnik, vol. 134, no. 3-4, 2017, pp. 287–289.

[6] L. El Asri and et al., "Frames: A corpus for adding memory to goal-oriented dialogue systems," in Proceedings of the SIGDIAL 2017 Conference, Association for Computational Linguistics, Ed. Stroudsbourg: Association for Computational Linguistics, 2017, pp. 207–219.

[7] K. Johnson, "Facebook messenger hits 100,000 bots," 2017. [Online]. Available: https://venturebeat.com/2017/04/18/facebook-messenger-hits-100000-bots/ 2017.08.28

[8] eMarketer, "Alexa, Say What?! Voice-Enabled Speaker Usage to grow Nearly 130% This Year: Amazon controls 70% of the voice-enabled speaker device market," 2017. [Online]. Available: https://www.emarketer.com/Article/Alexa-Say-What-Voice-Enabled-Speaker-Usage-Grow-Nearly-130-This-Year/1015812 2017.08.28

[9] Oracle and C. Parkes, "Can Virtual Experiences Replace Reality? The future role for humans in delivering customer experience," Redwood Shores. [Online]. Available: https://blogs.oracle.com/emeapartnercx/can-virtual-experiences-replace-reality 2017.08.28

[10] S. Quarteroni and S. Mandandhar, "A chatbot-based interactive question answering system," in Proceedings of the 11th Workshop on the Semantics and Pragmatics of Dialogue, R. Artstein and L. Vieu, Eds. Decalog, 2007, pp. 83–90.

[11] O. V. Deryugina, "Chatterbots," Scientific and Technical Information Processing, vol. 37, no. 2, 2010, pp. 143–147.

[12] S. Reshmi and K. Balakrishnan, "Implementation of an inquisitive chatbot for database supported knowledge bases," Sadhana, 2016, pp. 1173–1178.

[13] N. Dahlbaeck, A. Joensson, and L. Ahrenberg, "Wizard of oz studies - why and how," in Proceedings of the 1st international conference on Intelligent user interfaces, W. D. Gray, E. Hefley, and D. Murray, Eds. New York: ACM, 1993, pp. 193–200.

[14] F. Guerin, "Learning like a baby: A survey of artificial intelligence approaches," The Knowledge Engineering Review, vol. 24, no. 2, 2011, pp. 209–236.

[15] K. Panetta, "The ABC Technologies Will Change Future Customer Experience: Today is all about digital business, but CIOs should look beyond it to three transformative technologies," 2006. [Online]. Available: http://www.gartner.com/smarterwithgartner/the-abc-technologies-will-change-future-customer-experience/ 2017.08.28

[16] A. Deshpande, A. Shahane, D. Gadre, M. Deshpande, and P. M. Joshi, "A survey of various chatbot implementation techniques," International Journal of Computer Engineering and Applications, vol. 11, no. 6, 2017, p. unknown.

[17] N. Piyush, T. Choudhury, and P. Kumar, "Conversational commerce a new area of e-business," in SMART 2016, R. K. Dwivedi, Ed. Moradabad: Excel India Publishers, 2016, pp. 322–327.

[18] EY, "Listen closely: your customers are trying to text you something." [Online]. Available: http://www.ey.com/gl/en/services/advisory/ey-listen-closely-customers-are-trying-to-text-you 2017.08.29

[19] J. Bruner, "Chris Messina on conversational commerce: The O'Reilly Bots Podcast: The 2017 bot outlook with one of the field's early adopters," 2017. [Online]. Available: https://www.oreilly.com/ideas/chris-messina-on-conversational-commerce 2017.08.29

[20] C. Messina, "2016 will be the year of conversational commerce," 2016. [Online]. Available: https://medium.com/chris-messina/2016-will-be-the-year-of-conversational-commerce-1586e85e3991 2017.08.28

[21] A. Horzyk, S. Magierski, and G. Miklaszewski, "An intelligent internet shop-assistant recognizing a customer personality for improving man-machine interactions," Recent Advances in Intelligent Information Systems, 2009, pp. 13–26.

[22] E. C. Hirschman, "Predictors of self-projection, fantasy fulfillment, and escapism," Journal of Social Psychology, vol. 120, no. June, 1983, pp. 63–76.

[23] P. H. Bloch and M. L. Richins, "Shopping without purchase: An investigation of consumer browsing behavior," Advances in Consumer Research, vol. 10, 1983, pp. 389–393.

[24] J. R. J. Sherry, "A sociocultural analysis of a midwestern American flea market," Journal of Consumer Research, vol. 17, no. 4, 1990, pp. 13–30.

[25] E. Fischer and S. J. Arnold, "More than a labor of love: Gender roles and christmas shopping," Journal of Consumer Research, vol. 17, 1990, pp. 333–345.

[26] B. J. Babin, W. R. Darden, and M. Griffin, "Work and/or fun: Measuring hedonic and utilitarian shopping value," The Journal of Consumer Research, vol. 20, no. 4, 1994, pp. 644–656.

[27] J. Kanow and J. Earley, "The hedonistic paradox: Is homo economicus happier?" Journal of Public Economics, vol. 92, no. 1, 2008, pp. 1–33.

[28] A. F. Firat, N. Dholaki, and A. Venkatesh, "Marketing in a postmodern world," European Journal of Marketing, vol. 29, no. 1, 1995, pp. 40–56.

[29] B. Schmitt, Experimental Marketing: How to Get Customers to Sense, Feel, Think, Act, and Relate to Your Company and Brands. New York: Free Press, 1999.

[30] K. Scherer, "Emotions as a proceess: Functions, origin and regulation," Social science Information, vol. 21, no. 4/5, 1982, pp. 555–570.

[31] H.-J. Chang, M. Eckman, and R.-N. Yan, "Application of the stimulus-organism-response model to the retail environment: the role of hedonic motivation in impulse buying behavior," The International Review of Retail, Distribution and Consumer Research, vol. 21, no. 3, 2011, pp. 233–249.

[32] D. Kahneman, "Objective happiness," in Well-Being, D. Kahneman, E. Diener, and N. Schwarz, Eds. New York: Russell Sage Foundation, 1999, pp. 3–25.

[33] Kahneman, "Experienced utility and objective happiness: A moment-based approach," in Choices, Values and Frames, A. Tversky and D. Kahneman, Eds. Cambrigde: Cambridge University Press, 2000, pp. 673–692.

[34] N. Frijda, "The laws of emotion," American Psychologist, vol. 43, no. 5, 1988, pp. 349–358.

[35] A. Zainal, G. H. Sagala, and R. Muhamad, "The interchanges of hedonic-utilitarian values toward information system-users satisfaction: A case of social media users," Asia Pacific Journal of Contemporary Education and Comunication Technology, vol. 1, no. 1, 2015, pp. 167–178.

[36] M. R. Solomon, Consumer behavior: Buying, having and being. Upper Saddle River: Pearson and Prentice Hall, 2007.

[37] E. Cerezo, S. Baldassarri, and F. Seron, "Interactive agents for multimodal emotional user interaction," in Proc. of IADIS International Conference Interfaces and Human Computer Interaction, 2007, pp. 35–42.

[38] S. Etlinger and A. Altimeter, "The Conversational Business," Altimeter, a Prophet Company, Tech. Rep., 06 2017.

[39] R. Etemad-Sajadi and L. Ghachem, "The impact of hedonic and utilitarian value of online avatars on e-service quality," Computers in Human Behavior, vol. 52, 2015, pp. 81–86.

[40] P. Ekman and K. G. Heider, "The universality of a contempt expression: A replication," Motivation and Emotion, vol. 12, no. 3, 1972, pp. 303–308.

[41] C.-P. H. Ernst, J. Pfeiffer, and F. Rothlauf, "Hedonic and utilitarian motivations of socialnetwork site adoption," Working papers in Information Systems and Business Administration, 2013, pp. 1–14.

[42] M. Kaleem, O. Alobadi, J. O'Shea, and K. Crockett, "Framework for the formulation of metrics for conversational agent evaluation," in RE-WOCHAT, R. E. Banchs, R. Higashinaka, W. Minker, J. Mariani, and D. Traum, Eds., 2016, pp. 20–23.

[43] R. Eynon, C. Davies, and W. Holmes, "Supporting older adults in using technology for lifelong learning: The methodological and conceptual value of Wizard of Oz simulations," in Proceedings of the 8th International Confrence on Networked Learning, V. Hodgson, C. Jones, M. de Laat, D. McConnell, T. Rynerg, and P. Sloep, Eds. Lancaster: Lancaster University, 2012, pp. 66–74.

[44] L. D. Riek, "Wizard of Oz Studies in HRI: A Systematic Review and New Reporting Guidelines," Journal of Juman-Robot Interaction, vol. 1, no. 1, 2012, pp. 119–136.

[45] G. Buscher and R. Biedert, "Usability testing: Affective interfaces," Informatik-Spektrum, vol. 33, 2010, pp. 499–503.

[46] P. Ekman and W. Friesen, "Measuring facial movement," Environmental Psychology and Nonverbal Behavior, vol. 1, 1976, pp. 56–75.

[47] Microsoft., "Documentation: Emotion API," 2017. [Online]. Available: https://www.microsoft.com/cognitive-services/en-us/emotion-api/documentation 2017.09.12

[48] Microsoft, "Cognitive Services: Emotion API," 2017. [Online]. Available: https://westus.dev.cognitive.microsoft.com/docs/services/5639d931ca73-072154c1ce89/operations/563b31ea778daf121cc3a5fa 2017.09.12

[49] M. Kearn, "The many features of Cognitive Servies - Part 1 - Vision," 2016. [Online]. Available: ttps://blogs.msdn.microsoft.com/martinkearn/2016/ 06/06/the-many-features-of-cognitive-services-part-1-vision/ 2017.09.14

[50] Arrowchat, "ArrowChat: JQery/PHP Chat Software and Facebook chat plugin for websites," 2017. [Online]. Available: https://www.arrowchat.com/ 2017.08.31

[51] B. Rammstedt, C. J. Kemper, M. C. Klein, C. Beierlein, and A. Lovaleva, "A short scale for assessing the big five dimensions of personality," Methoden, Daten, Analysen, vol. 7, no. 2, 2013, pp. 233–249.